The Recycling Myth

The Recycling Myth

DISRUPTIVE INNOVATION TO IMPROVE THE ENVIRONMENT

Jack Buffington

 PRAEGER™

An Imprint of ABC-CLIO, LLC

Santa Barbara, California • Denver, Colorado

Library of Congress Cataloging-in-Publication Data

Buffington, Jack.
   The recycling myth : disruptive innovation to improve the environment / Jack Buffington.
      pages cm
   Includes bibliographical references.
   ISBN 978-1-4408-4307-5 (hardback) — ISBN 978-1-4408-4308-2 (e-book)
1. Recycling industry.   2. Business logistics—Technological innovations.   I. Title.
   HD9980.5.B798 2016
   363.72'82—dc23   2015030218

ISBN: 978-1-4408-4307-5
EISBN: 978-1-4408-4308-2

20 19 18 17 16   1 2 3 4 5

This book is also available on the World Wide Web as an eBook.
Visit www.abc-clio.com for details.

Praeger
An Imprint of ABC-CLIO, LLC

ABC-CLIO, LLC
130 Cremona Drive, P.O. Box 1911
Santa Barbara, California 93116-1911

This book is printed on acid-free paper ∞

Manufactured in the United States of America

Contents

1. The Myth   1

2. Frankenstein's Bottle: A Problem of Math   17

3. The Throwaway Supply Chain   39

4. The Happy Cup Fallacy   53

5. Bottles Grown from the Soil   67

6. It's in Our Blood   83

7. Something Needs to Be Done!   103

8. Solutions—Extending the Supply Chain to Nature   113

9. 2020: The Happy Cup Reality   137

Notes   143

Index   155

Chapter 1

The Myth

Americans on average use three beverage containers a day, from a disposable can, plastic or glass bottle, plastic-lined paper cup, to other assorted container options (Table 1.1). They thus use more portable, throwaway beverage containers per capita and volume than anyone else in the world. With a U.S. population of 314 million, this figure adds up to approximately 1 billion cups, cans, bottles, and other containers being shipped, sold, processed, recycled, and landfilled in America on a single day! Consuming beverages is a second-nature activity for Americans; we are never more than a few minutes away from grabbing a drink from our refrigerator or stopping by a coffee shop or convenience store. Can we imagine what our lives would be like if hot and cold drinks weren't so portable and readily available at a reasonable price? How many of us actually think about the supply chain system that is required to offer such convenience and cost affordability? Not many, I'm afraid to say. Between 2000 and 2010, beverage container use in the United States grew by 22%, but the recycling rate fell from 41% to 37%,[1] and this statistic does not include the beverage containers for which there is no attempt to recycle. One billion containers used in a day, and at best, only 300 million recycled, with many of them eventually thrown back into the trash, exported, or used in a lesser purpose. Today, the U.S. packaging industry has revenue of over $170 billion annually,[2] a big business that, intentionally or not, is largely driven by waste.

If you ask most anyone associated with the consumer packaging industry—someone responsible for making or filling the container, a waste management industry specialist, or even an environmentalist—the solution you will hear is this: Americans and others around the world

Table 1.1   Per Capita Portable Beverage Container Per Day, U.S.

Material	Use Per Capita (U.S.)
Aluminum Can[a]	0.88578
HDPE Plastic[a]	0.07529
PET Plastic[a]	0.69977
Coffee Cups (Plastic-Lined Cardboard)[b]	0.22144
K-Cups (Single Serve Coffee)[c]	0.07972
Takeout Plastic Cup[d]	0.88578
Other[a]	0.17716
TOTAL	3.02494

Source: Container Recycling Institute,[a] Carryyourcup.org,[b] Hamblin, 2015,[c] and estimate.[d]

should "recycle more." To many in the United States and around the world, the answer is to implement a *refund-deposit system,* or *bottle bill* program. Mandated bottle bill programs will, by definition, lead to higher collection rates, and that means fewer beverage containers are sent to a landfill after consumer use. In some European nations, such as Germany and Sweden, collection rates are over 90%, leading to a near *zero waste* society where nothing is sent to a landfill. The popular media in the United States has embraced this approach for eliminating packaging waste and its impact on the environment: enact a nationwide mandatory refund-deposit program like exists in Europe, and the United States can become a *zero waste society* as well. However, from this book you will learn that, not only are such *zero waste* countries not what you may envision, these legislative programs with the best of intentions may actually be counterproductive to achieving economic and environmental success in the long run.

The *recycling myth* is the perspective that is preached by industry, government, advocacy groups and even consumers—that if we all *do our part* and recycle, this problem of litter and packaging waste can be eliminated. However, this book will show that the act of recycling is ineffective today because it has a goal to *do less harm* to the environment rather than *improving the environment*; essentially, it is a waste management mitigation scheme rather than the use of innovation to optimize both economic and environmental factors. Promoting a solution to do less harm causes an opportunity cost in failing to do something better and actually improve the environment. The goal of this book is to bring about an awakening that we must do a lot more than collect a beverage container to solve the problem of waste; we must enable an innovation engine toward science, technology, and supply chain logistics for real solutions to achieve a paradigm shift in consumer products. This I know for sure because, not only am I a supply chain leader for a top United States beverage company, I am also a post-doctoral researcher in supply chain and biotechnology at the Royal Institute of Technology in Stockholm, Sweden—a top university in a country known for its environmental practices. The environmentally conscious Swedes are light years ahead of Americans in regard to their recycling adherence and acumen, but their solutions will, at best, mitigate rather than solve the problem of packaging waste and environmental impact. There's more to the problem of packaging waste than the collecting of something that has little economic value.

In the United States alone, 1 billion beverage containers are used in a single day, of which 700 million containers are thrown into landfills or incinerated as waste. Included in this waste total are materials (primarily aluminum) discarded that actually are in market demand if collected. Of the 300 million containers that are recycled daily, millions are eventually discarded into a landfill anyway after going through the whole recycling effort; this is due to the material being contaminated or not at the level of quality required, or even due to a lack of demand in the secondary reuse market. According to Susan Collins from the Container Recycling Institute, about a quarter of all materials recycled using a single stream curbside program (everything thrown into one bin) is eventually thrown away, and 40% of glass packaging.[3] Of the remaining 70–75% of recycled materials, virtually all recycled aluminum is reused for like purposes in packaging, building, and transportation, while most glass and plastic is only used for *downcycled* (lesser) purposes in order to avoid landfilling. Mandatory recycling practices, like a bottle bill program, may appear to be a more environmentally friendly waste management strategy than landfilling but does not incentivize innovation to solve the problem. If modeled as nature, an optimal recycle/reuse program would ensure that all discarded containers are reused synergistically in a circular ecosystem; there are no waste management programs in nature! Calling out this recycling myth is not to promote doing nothing but rather to spur a materials revolution and supply chain practices that create symbiosis, not forced practices. The problem needs to be solved, not mitigated or legislated.

The recycling myth is the story of how our current environmental practices are unintentionally doing more harm than good, and how we need to create a supply chain of the future that will design, produce, consume, and reuse our materials in a manner that is balanced economically and environmentally. Nowhere in nature are there mandated recycling programs; organisms are not acculturated to dispose or incinerate its residual excess, rather it is just left behind in order to be a part of another set of activities within the ecosystem. Of course, nowhere in our ecosystems are synthetic materials either, so there is no need for getting rid of foreign materials, like plastics, that can cause undue stress on the environment, wildlife, or our own bodies, if left unchecked. We must realize that an after-the-fact recycling strategy as a back-end supply chain practice can never be effective when materials are designed poorly in the first place. While studying in Sweden, one of the most environmentally friendly places on earth, I learned that even the most perseverant and dedicated back-end recycling programs can only minimize the damage to both the economy and environment when a material is poorly designed, such as the plastic bottle. I applaud the efforts from environmentally astute cultures and those in the United States who advocate recycling, but I wonder if these efforts unintentionally take the pressure off poor product design and supply chain systems in the end.

The math of America's beverage consumer behavior is as follows: in 1970, as many beverage containers were recycled as wasted, but in 2010, the amount wasted was approximately 50% higher than recycled, despite decades of environmental policy and programs.[4]

Table 1.2 demonstrates our conventional approach to recycling as a losing battle: in reality, our consumer products supply chain has been designed for front-end growth as mutually exclusive from reuse efficient, leading

Table 1.2   Mountains of Waste

Year	MSW Waste Generation (U.S.) (millions tons)	Per Capita Waste Generation (lbs/person/day)
1960	88.1	2.68
1970	121.1	3.25
1980	151.6	3.66
1990	205.2	4.50
2000	239.1	4.65
2010	250.6	4.44
2013	254.1	4.40

Source: xfuels.com.

to growing waste. This is happening worldwide, especially in developing nations where consumerism is ramping up much faster than an effective waste management system can handle. In the fast-growing, low-recycling economies of Asia, to the mature, high-consumption, moderate-recycling nation of the United States, to more environmentally conscious nations in Europe, the problem is mitigated at best and a crisis at worst. Indicative of the problem, the back-end recycling sector in the United States is largely decentralized, small scale, inefficient, and decoupled from mainstream supply chain systems. Therefore it is often unable to integrate into a combined cradle-to-cradle system (i.e., a biomimetic system modeled in nature's processes—sometimes referred to as C2C) with the massive primary material providers of plastic, glass, and aluminum. Yet in the marketing of recycling programs, there is the perception that a circular supply chain is in place, similar to what exists in a natural ecosystem; metaphors of recycling as a closed loop system where materials are collected and reused, like for like, is more preached than practiced. The chasing arrows on your beverage container are more hope than reality.

In chapter 2, I classify Nature as *Material Scientist 1.0*—able to achieve these closed-loop ecosystems—in comparison to Man as *Material Scientist 2.0*—who develops synthetic materials like the plastic water bottle that Nature would never design and manufacture and, therefore, cannot reuse. Truthfully, most disposable beverage containers have been designed to optimize only industrial supply chain properties for consumer preference, such as overall strength, tensility, light weight, and cost, with little consideration that the product designs are foreign materials to nature that are unable to be decomposed and reused as a participant in nature's closed-loop system. Of course, this disconnect should not be surprising: in an emergent environment of adaptability and self-organization, Nature's materials grow up with each other, so to speak, over hundreds of thousands, if not millions of years. In contrast, our industrial production system (especially the system for creating bottles, cups, and cans) was designed with the opposite principle in mind: we design synthetic materials to be *nature proof*—resistant to the hunger of creatures doing their part in decomposition, to contamination, disease, heat transfer, and damage. Synthetic materials are superior in balancing the lightweight features of packaging at the highest strengths and oxygen and microbial protection at the lowest cost to ensure the integrity of the beverage. They have broken the code of nature's closed-loop system, which we now realize comes with unintended consequences. You may think that disposable coffee cups are better because at least they're

partially made of paper—wood pulp—a biodegradable material. However, precisely because plastic was introduced into the design to keep the
heat of your drink inside the cup and away from your hands, they don't
decompose. The technological innovations that gave us these very useful containers took a shortcut from nature, building solutions without
understanding their implications for the long run.

In this industrial ecosystem, we can produce a massive array of throwaway containers at very low costs, and on the back end, it is more efficient to bury these materials in our industrial landfills than it is to recycle
them. Consumers are told it is efficient to reuse when in reality, it is
not. This is not due to deceptive practices that the consumer beverage
industry is sometimes accused of, but rather the nature of how the industrial supply chain system works in practice. In this industrial process,
our supply chain systems go through astounding heights of efficiency
in front-end processes of extraction and manufacturing in order to produce cheap materials that are buried or burned after thirty minutes of
use. This is drastically different from when our industrial supply chains
were less efficient, given a scarcity of resources and the high cost of production and distribution, naturally leading to recycling and reuse programs. Back then, the front-end and back-end systems were mirrored in
efficiency (or inefficiency). Given the hyper-efficiency of today's supply
chains, we have reached the dreadful point where it is actually more
efficient to landfill or incinerate a resource than to reuse it.

Therefore, the root cause must be addressed long before we ask a
consumer to put her container into a magic bin; we must fix this fracture
between our natural and industrial systems. The nature system reaches
a balance between use and reuse through organic materials while the
industrial system has been effective in nature-proofing the process
through synthetic, unnatural materials that are cheap, strong, lightweight, and sanitary. This leads to a counterproductive fracture between
the objectives of the environmentalist, who wishes to save the planet
with the industrialist, who seeks to maintain economic growth. Without
a healthy balance between economic and environmental objectives, both
will suffer; an out-of-balance model that focuses solely on consumption
and waste or nature and preservation is the ultimate lose-lose situation.

Americans must also acknowledge that our current economic growth
paradigm is built upon the use of throwaway containers. Marketing programs and technology push innovation in consumption and production
but not in reuse. Unfortunately, our mega supply chains have been built to
optimize using primary ("virgin") materials rather than reusing a majority

of secondary materials within these economies of scale systems. Recycled materials are inefficient to the overall supply chain and manufacturing process, with smaller, less efficient recycled material providers unable to achieve similar relationships with manufacturers as large scale sectors, such as aluminum and petrochemical. Unfortunately, this leads to materials from recycling programs being outside the scope of the economy of scale consumer product supply chains, leading to a conflict of whether recycling is better because it diminishes waste or worse because it leads to inefficiency. As long as natural resources such as petroleum and aluminum are priced on the open market at lower prices than recycled material, beverage container manufacturers in ultra-competitive markets will choose primary materials via its supplier relationships. Environmental regulation, such as mandatory recycling and extended producer responsibility programs, can change this equation, but not without disturbing the current economic price structure that the consumer in the United States demands; in Europe, there is a culture of less consumption and higher pricing that may not be seen as a welcome trade-off in the United States.

ALICE IN RECYCLING-LAND

Despite the mounting evidence of the irreparable ecological damage occurring to the environment as a result of disposable packaging, particularly our oceans, there are those who mock the notion that anything needs to be done in the United States and other nations. According to the late, great comedian George Carlin, our anthropogenic impact on the earth is no problem. In a live skit, he told the audience that "humans are so self-important . . . we haven't learned how to take care of ourselves, and now we are going to save the (expletive) planet. . . . The earth doesn't share our prejudice against plastic; plastic is just another one of its children." Whether a tongue-in-cheek comment or not, there are too many, at least in the United States, who doubt that the massive introduction of synthetic materials into our ecosystems is an important problem. Regardless of what one thinks of today's environmentalist ideologies, I wonder how anyone can justify the use of a resource just once without considering a need to reuse it on a planet of finite resources and a growing world population on its way to ten billion people. This is truly Alice's upside-down world: how can a mind-set of apathy exist regarding 700 million beverage containers a day for burial, incineration, or export shipping to developing countries because most of these containers were never designed to be reused? Even if the U.S. waste management system is operating at over 99% efficiency, this would lead to at least 6 million containers escaping the system daily and entering the ecosphere where it can end in our oceans, waterways, other natural ecosystems, or even our own bodies. Of course, in the developing nations of Asia, particularly China, more packaging waste is being created at lower efficiency rates, leading to an astronomical negative impact to the environment. Inherently, a back-end environmental mitigation approach to a front-end consumer products issue becomes a collection of bad science and bad economics, leading to false choices and a growing ecological problem that will negatively impact the global economy as well. No doubt George Carlin was funny, but wrong!

Prior to the Industrial Revolution, a different waste management problem existed: New York had a population of 1.2 million in 1880 (in contrast with 8.5 million today), and its biggest problem was preventing disease. Unlike today's modern world, our cities weren't synthetic and sanitary. In contrast to today's New York City of shimmering glass and steel towers and a super-efficient supply chain system of sanitary, synthetic consumer products and waste management, the city of 1880 was an unsavory melting pot of disease, including family livestock and an army of horses used as transportation. The family goats, pigs, and chickens were problems, but horses presented the greatest threat to health, with 150,000 horses producing 45,000 tons of manure a month.[5] Back then, waste removal was a matter of life and death; in contrast, today's waste management strategy is one of scale and convenience of largely sanitary waste rather than human disease from filth. Eventually, the Industrial Revolution led to great innovations, most notably, the use of fossil fuels that made New York City and the rest of America powered by mechanical machines that belched CO_2 rather than crap manure. The industrial supply chain systems gave us freshly managed food and drink, eliminating the dangers of raising livestock in close quarters. This scalable, sanitary, and portable industrial supply chain system became so efficient that, over time, it would lead to an upside down definition of optimization that would lead to us throwing away billions of one-time use articles worldwide that is turning our organic planet into, as Pope Francis noted, "an immense pile of filth."

Eventually, the industrialized world would have to pay greater attention to this growing mountain of sanitized waste; in 1987, a massive trash crisis took place in New York City. The jurisdiction of Long Island was practically out of landfill space and, as a result, could no longer send its school and commercial trash to the Islip, New York, landfill for processing. The solution was to ship the trash that could not be buried in its landfill to less populated regions north of the city but at very high prices. Clearly, the city was running out of options and was desperate to try any solution. Then, the city received a unique proposal from a good ol' boy, serial entrepreneur from Alabama and a notorious gangster with a stranglehold over the trash business in New York City. Alabaman Lowell Harrelson began his career in the construction business in the 1970s, made a fortune, and then lost most of it during the oil collapse of the early 1980s. His next idea was even bigger than taking advantage of a building boom in the South: to address America's burgeoning trash problem through an innovative waste-to-energy program that took tightly baled trash from heavily populated areas to rural southern farms in order to create landfills that would also produce methane to be used for energy capture. This was the type of big idea that could save New York City from being buried alive in its own trash, but first Harrelson needed a business partner who could open doors for him.

Anyone who understands the waste management business in New York City in the 1970s and '80s knows the business was owned by the mob, headed by the Lucchese crime family. During this period, the mob had a near monopoly, using intimidation and even murder to tighten its control over competition. On Long Island, family captain Salvatore Avellino owned the business for 15 years. Harrelson's brilliant idea and Avellino's connections brought them together to pitch a solution to desperate Long Island officials. On paper, the business pitch looked great. The two men offered to take the trash from the city for $50 a ton versus the $86 they were currently paying to haul trash to northern landfills.[6] The two entrepreneurs planned to save money by hauling the trash to North Carolina on a barge, a cheaper form of transportation than trucks, and then sending to a landfill site at $5–10 a ton.[7] These trash farms would receive dumping fees as well as earn revenue from methane production from the waste. Harrelson and Avellino would have become waste management innovators if not for one thing: America's phobia of trash.

The desperate Long Island officials, Harrelson, Avellino, his business partners, and six adjoining landowners in North Carolina were brought together in a hasty agreement. The maiden voyage consisted of two tugs (the *Break of Dawn* and the *Mobro 4000*) that would haul trash from Long Island to a North Carolinian port, and then to the new landfill sites. The *Mobro 4000's* first load was 3,186 tons of trash tightly baled to keep it secure.[8] As the barges set out to sea, one can only imagine how

Lowell Harrelson and Salvatore Avellino dreamed of getting rich and becoming America's waste management heroes, but what happened was quite a bit different from their wishes. Rumors of the barges that were steaming toward North Carolina—based upon a handshake deal between a businessman from Alabama, New York mobsters, and North Carolina farmers—raised suspicions among local officials and the media, which soon turned concerns that there were radioactive materials hidden in the trash, or even some dismembered enemy of the mob! Before the barge reached North Carolina, the public pressure became too great, and the barge was not admitted for entry. Caught completely off guard, Harrelson had to find another home for the trash before the media spread the news much further—or before the mobsters began to lose patience. Perhaps there were small skirmishes between trash businessmen and environmentalists in various local communities around the country at the time, but this one took a national, if not international, media glare, raising awareness of a significant problem. As a result, Lowell Harrelson became the unintended poster child of the modern recycling movement, instead of rich entrepreneurs and environmental innovators.

In 1987, America woke up to its trash problem through the story of a floating rubbish barge with nowhere to dump; it was on the nightly news across the nation and a Johnny Carson late-night punch line. The environmentalist movement used the media attention as an impetus for promoting America's need to recycle. Lowell Harrelson's naiveté in doing business with the mob without formal approval from local officials was an unintended catalyst to the movement, although he saw it much differently: "I still cannot see the flaws in the plan, except for the psychological problem that people have with garbage."[9] However, what was inconceivable in 1987 is commonplace today: New York's trash is sent to other states, such as Pennsylvania, Virginia, and even South Carolina. Also, New York City is making $12 million a year from methane capture.[10] In 1987, America suddenly became aware of its trash problem as a waste management concern. Today, the same problem exists; even if it is not as visible to us, our problem of throwaway consumer products and waste is increasingly urgent. Unfortunately though, we are focused on the symptoms rather than the problem, and people like Lowell Harrelson become the scapegoat, preventing us from addressing the real issue.

Almost thirty years later, we continue to chase symptoms and patch Band-Aids upon them through recycling programs that will never solve the problem. America's array of mega supply chains of waste management whisks away the litter and then hermetically seals most of it within large tombs in its vast landmass; in Sweden, there is a culture of conformance to tidily sort, clean, transport, and reuse materials. If this cannot be done, then the materials are incinerated in a waste-to-energy scheme in order to achieve a near zero waste status, at least in concept. Current recycling programs and, for that matter, modern waste management disposal practices are akin to putting a finger in the dike; it slows the pace of the collapse but never fixes the problem. Solving the problem is not asking the consumer to consume less, the producer to reduce production and process materials largely unwanted, or a treatment facility to burn or bury more material. Instead, we need materials within a supply chain system like molecules and organisms within an ecosystem; those involved contribute positively in relation to the entire system.

THE 411 & 911 OF TRASH (AND RECYCLING)

What has been accomplished since 1987 is this: due to the efficiency of our mega waste management supply chain system, America's trash is largely invisible, but this wasn't always the case. As a kid in Baltimore in the 1970s, I knew the local dump as an eyesore to the community, an unsightly, disorganized farm of trash, requiring heavy equipment and a manual workforce. Today, there are few local dumps, but primarily a hub-and-spoke mega supply chain system consolidated into a big business, its landfills moved to sites far from where most people live. If you venture out and tour one of these super sites, you'll see a well-designed facility, possibly including a power generation plant and maybe even trees and wildlife on its border areas. Just as the little mom-and-pop corner stores of yesteryear were replaced with Wal-Mart super stores, trash removal has also consolidated. Today, nearly 47% of all U.S. waste is handled by two companies: Waste Management and Republic Services, yet 39% of the industry remains with small companies with less than 1% market share.[11] In 1992, private industry managed 35% of waste, and in 2012, it was 78%, a further demonstration of the commoditization and privatization of the waste management process as big business.[12]

In contrast to today's mega landfills and waste management companies, recycling operations remain small, decentralized, inefficient, and disconnected from the front-end industrial supply chain. As a result, the recycling sector cannot compete with a dominant design supply chain that primarily relies on virgin materials of petrochemicals, lumber, and aluminum feedstock that is viewed as optimal. Add to the storyline that most of today's beverage containers were not designed to be reused in the first place, it makes little difference in the United States whether you recycle your plastic water bottle or throw it in a trash bin. Most waste management companies offer curbside recycling services for a fee, but we are left to wonder if this is more marketing than progress since they make higher profits on the trash side of the business; in 2015, Waste Management lost $16 million in the first quarter in their recycling division, a staggering indicator that this business model is yet to be profitable.[13] In its final calculation, the collection, recycling, and reuse of materials is less efficient than throwing it all into a sanitary hole, as troubling as that may be to conceive.

With recycling programs being small in scale in comparison to the mega supply chains of consumer products and waste management, and therefore inefficient, its raison d'être is largely enabled by government regulation and environmental advocacy. In 1987, the EPA and states set recycling goals, and nationwide curbside programs grew from 1,000 in 1988 to over 7,000 in 1995.[14] Various recycling programs such as drop-off centers, "pay as you throw," and buy back recycling programs and bottle bills took shape as efforts to redirect materials from landfills to reuse. It has been proven that mandated programs lead to higher recycling rates, but it should not be assumed that higher collection rates lead to higher secondary material reuse. In reality, collecting more of something that cannot be efficiently reused does not solve the fundamental problem but only mitigates the mess, at best.

JUST MAKE IT GO AWAY!

Today, each American man, woman, and child produces 4.38 pounds of trash a day,[15] including 3 beverage containers; this translates into roughly 250 million tons of trash a year, which is a 283% increase from 1960.[16] If not for the technological engineering marvels of our modern day waste management system, there would be significant economic and environmental chaos preventing our current levels of consumption or leading to more government intervention. Without such trash challenges, we are only left with the psychological guilt of not recycling, a mere shadow of a concern that many of us think very little about. The question is this: Does recycling matter when our waste management techniques are so efficient, and if so, it is making a difference? And if recycling is not making a difference in its approach, what is the alternative?

In this book, I will present a mountain of evidence to show that today's conventional approach to recycling is not, and can never be, successful. This is a sticky issue, however. In my research, I have been criticized for challenging the ineffectiveness of recycling programs, given its sacred-cow status in the realm of environmentalism; while not considered perfect, its advocates believe that "it's better than nothing," as if nothing is the only alternative. In this book, I make a case for how to improve both the economy and the environment by creating packaging materials and other consumer products that are designed for good to transform how we live in the industrial and natural 21st-century system. My case will go beyond a theoretical model, with potential solutions that could be achieved in less than a decade. This is absolutely necessary in the United States and other nations, given the words of Waste Management executive David Steiner: "If people feel that recycling is important—and I think they do, increasingly—then we are talking about a nationwide crisis."[17]

I will start my case for why conventional recycling programs cannot work by introducing "Frankenstein's Bottle" in chapter 2. This is the story of Nature as *Material Scientist 1.0*, followed by Man as *Material Scientist 2.0*. Despite Man's grand efforts to play God and invent his own materials, something has gone horribly wrong through the creation of plastic bottles, glass bottles, plastic cups, and plastic-lined paper cups. Nature, as a designer, defined ecological balance at a glacial pace over tens, if not hundreds, of millions of years, achieving cohesiveness between organisms in an emergent natural system. Man as *Material Scientist 2.0* cannot show such patience, inventing materials that would achieve industrial growth without consideration of how these materials would correspond with nature, or even repel it. As a result, unparalleled economic growth has been achieved in the United States, with the collateral damage of growing waste, which is just the tip of the iceberg when one considers the filth growing in the developing world. Essentially, we and our environment have become victim to our poor definition of design and efficiency.

In the United States, the packaging waste problem is enabled through a throwaway society, which I discuss in chapter 3. How did we get to this point? When I was growing up in the 1970s and '80s, I was taught that "waste is a sin" and the problems of a "material out of place." Yet in the 20th century, there was a need to achieve economic growth in order to maintain high standards of living in the United States, even to the point of promoting obsolescence and waste. A culture of massive waste was not desired, but rather, was seen as a de facto requirement in order to maintain economic growth after World War II. The bad news is that we now have generations of Americans who know nothing other than such waste and one-time throwaway misuse. The good news is that in the 21st century, we now possess tools that can help us balance economic growth and environmental sustainability without waste.

Chapter 4 presents the concept of the "happy cup" fallacy and how this marketing concept allows us to be content with an exorbitant amount of economic and environmental waste that each of us creates every day. Starting in the 1980s and continuing today, America's love affair with throwaway packaging ascended to the current rate of 3 disposable beverages a day. To counteract the negativity of personal guilt from so much waste, sustainability programs were introduced to manage behavior—to encourage both recycling and consumption at the same time without reconciling the obvious contradictions. Peter Senge's term, the "happy cup" fallacy, defines this phenomenon of consumer guilt being washed away through the intertwined terms "recyclable," "biodegradable," and "compostable." Once their containers are whisked away, a consumer does not know, or even care, whether these articles are actually recycled or reused, but they are marketed to imply that recycling means bottle to bottle, can to can, or cup to cup reuse. In reality, it is the efficiencies of our modern day waste management system, not recycling, that enables consumerism by burying the problem into sanitary graves. Is this the sweeping of the dirt underneath the rug instead of confronting the problem head on? If we are able to design our materials in order to do good rather than do less harm, we are no longer conflicted and the need for a happy cup fallacy goes away.

Chapter 5 discusses another sustainability phenomenon and its impact on beverage containers: the $35 billion and growing U.S. organics industry. Food is just one aspect of organics; it has moved into cosmetics, clothing, and bottles and cups as well. Plant-based bottles are becoming all the rage, with big players like Coke pumping millions of dollars into organic bio-material bottle initiatives. Is this real sustainability or more of the same happy cup fallacy? At the present, these biomaterials are simply an extension of conventional packaging, meaning that it is achieving marginal improvements in recycling and sustainability, despite the message being sent to the consumer. The general approval of an organic bottle demonstrates the fallacy that if something is organic, it's always better for the earth. Unfortunately, this can become a distraction from real disruptive innovation that is absolutely necessary to address environmental and economic problems.

Pelle Hjalmarsson, the chief executive of Returpack, the recycling arm of Sweden, has been quoted as saying that Swedes are so passionate about recycling that "it is in our blood to make deposits." Environmentalists hear these remarks and congratulate the Swedes for being so much more advanced than Americans when it comes to sustainability. But are they really? In chapter 6, I debunk this recycling myth for two reasons: First, what works for 9 million Swedes doesn't necessarily work for 314 million Americans. Second, and more important, the Swedish sustainability programs are not solutions, but rather a large bandage. In addition, Swedes, particularly the younger generations, are showing signs of a growing recycling fatigue, and I suggest that they are less enamored with their own definitions of success than Americans perceive. Two questions stem from this chapter: Should Americans and others traipse down a path of recycling that mirrors the Swedes, which may not be as advertised? Would the average Swede continue to adhere to such laborious processes if it were understood how ineffective they are? Perhaps what's in their blood needs to change.

How will we move past this recycling myth and into the world of 21st-century solutions for economic growth and improving the environment? It is my belief that we head in the direction of improvements based on material science innovation (e.g., *green chemistry*) and design and supply chain transformation. Then we build the appropriate collection and reuse procedures. But what if instead the United States sought to impose a nationwide mandated recycling program, like in Sweden? I attempt to model what would happen in chapter 7. Building from chapter 7, chapter 8 provides a different approach to solving the packaging waste problem in the United States. In this chapter, I lay out the five components that are required in order to balance both economic growth and environmental sustainability. In creating this balance, an entirely new approach to recycling and environmentalism is set forth, driven by disruptive innovation. Business begins to model nature and leads to dramatic and reformative change; a new entity takes responsibility for designing our packaging material! Finally, a balance is achieved that enables today's diametrically opposed stakeholders (such as private companies and environmentalists) to work together on common goals. This chapter will provide new insight to the reader that problems can be solved through balance and collaboration, not contention.

In an interview prior to leaving office in 2013, outgoing Nestlé Waters North America CEO Kim Jeffery noted, "We have a broken system of recycling in America. Nobody is winning right now on this thing. We're not moving the needle."[18] This crisis of waste is something that we can no longer ignore, even if it is unintentionally enabled through the super-efficiency of our waste management system in the United States, or recycling mitigation schemes in Sweden. Waste is not just an environmental problem, but an economic one as well; trading one for the other simply creates a new problem to address. Either being buried in a hole, or incinerated while counted as being reused, such ploys are nowhere near Nature's process of sustainability. In the last century, two great and diametrically opposed errors were made that led to an impasse: one, that economic growth could only be forged through additional waste via throwaway, one-time items, and two, that mandated recycling programs and extended producer responsibility could mitigate this waste. Both of these fallacies exist as bedrocks of bipolar conventional thinking, yet in the 21st century, there does not appear to be a good reason for isolating each through incremental innovations on both sides; with a willingness to consider disruption to both industry and environmentalists, let us begin the journey.

Chapter 2

Frankenstein's Bottle:
A Problem of Math

WHO'S THE MONSTER?

During a wet summer on Lake Geneva in the year 1816, three great minds gathered in a remote cottage, sharing the depth of their collective wisdom. Two of the finest poets England has had to offer, Lord Bryon and Percy Shelley, and an unknown 19-year-old romantic interest of Shelley's named Mary Godwin, challenged each other's intellect at the rustic hideaway, far from the bustle and noise of old London. One night around the fireplace, Lord Bryon proposed a "ghost story competition"; the two men quickly became bored with the idea while Mary was determined but experienced writer's block for a few days. One night after a deep conversation regarding the meaning of life, Mary Godwin Shelley had a "waking dream," and developed a short story of a corpse being "galvanized" to life through the efforts of a young scientist named Viktor Frankenstein. This first-ever piece of science fiction literature was developed into a novel in the year 1818 under the name *Frankenstein; or The Modern Prometheus;* at first, it was published anonymously and finally under the author's name, Mary Shelley.

Nearly two hundred years later, this novel continues to thrill us with the limits of science and human progress, the role of technology in society, and the notion of humans "playing God." During the Industrial Era, humanity has been enamored with the ability to invent and build a world of its own, purposely separate from the natural world. In this chapter, I discuss how the history of industrial material design has led to a model in beverage packaging in which the math makes sense for the creation of cups, bottles, and cans to be used for thirty minutes and then become waste in the environment for an eternity. Can efforts to reverse engineer something be successful when it was never designed to be reused in the first place? The focus of this chapter is not the back-end recycling system, but rather the front-end material design process of packaging containers in which synthetic materials were manufactured from nature's warehouse before they were understood and are now being used and wasted by billions worldwide on a daily basis.

According to GreenBiz (shown in Table 2.1), the value of wasted packaging material in the United States is estimated as $11.4 trillion, or 68.3% of the annual U.S. GDP.[1] This leads to the question: If these used beverage containers, or UBC, are so valuable, then why are they buried and incinerated, or, not trash-mined afterward? To answer this economic paradox, I will tell the story of these materials and how Man as Material Designer 2.0 came into conflict with Nature as Material Designer 1.0 due to a problem of math.

Many of today's synthetic materials, such as plastic, are designed to be nature proof and therefore must be classified as a misapplication of science. There are no emergent, self-organizing mechanisms in place with these materials and the natural world, so inevitably, it becomes (or will become) toxic waste when discarded in a landfill or when it escapes into the oceans. I call these beverage containers *Frankenstein bottles* for good reason: we have been producing these synthetic creations in mass quantities

Table 2.1   Value of Wasted Packaging Materials in the United States[2]

Material	Estimated Value ($)
Paper (Coffee Cups)	$1.3 trillion
Glass (Beer and Salad Dressing Bottles)	$97.3 billion
Steel (Soup, Food Cans)	$285 billion
Aluminum (Beer and Soda Cans)	$1.5 trillion
PET Plastic (Water Bottles)	$2.9 trillion
HDPE Plastic (Milk Jugs)	$2.9 trillion
PVC Plastic (Cleaner Bottles)	$136 billion
LDPE/LLDPE Plastic (Chip Bags)	$726 billion
PP Plastic (Mayo Jars)	$1.3 billion
PP (Takeout Containers)	$371 billion
TOTAL ESTIMATED VALUE	$11.4 trillion

for a few decades without sufficient consideration of their long-term impact on the environment and economy. Worse, we do have some initial unsettling conclusions: one recent study has found that 4 to 12 trillion tons of plastic are washed into our oceans annually, which is equivalent to 1.5–4.5% of the total worldwide production.[3] Large patches of our oceans are plastic soup, some regions the size of Texas, possessing a greater concentration of plastic than phytoplankton, the life blood of the marine ecosystem. Even places far from civilization, expected to be untouched by human hands, such as the Arctic Ocean, have large concentrations of plastic frozen into the sea ice.[4] One does not need to be a marine biologist to understand that if our oceans die, this will eventually impact us economically and threaten our very existence. Yet the full extent of this damage remains to be seen, as our production, use, and waste of synthetic materials continues to escalate.

The creation of these synthetic materials began in the 18th century with humanity's mastery of chemistry; bad applications of science led to bad design, which drives the utter hopelessness of recycling, leading to trillions of dollars of resources being sent to landfills and incinerators, as shown in Table 2.1. In contrast to Material Scientist 2.0's processes, Nature as Material Scientist 1.0 almost always uses an efficient set of ingredients, processes, and relationships to create materials. Man's understanding of Nature's chemistry is just starting to be realized, as Nature's nanoworld of design is seen through technology and revealed as a beautiful compilation that has taken place over billions of years, leading to the symbiotic ecosystems that we appreciate today. The more we learn of the intricacies of nature, these temporal computations, the more we understand how limited our view of design is in comparison. Man's functional understanding of the sciences, such as chemistry, has led to wonderful innovations in material science but is mathematically inferior to the eternal patience found in the symbiotic relationships formed over thousands, if not millions, of years on earth. Yet if Industrial Man is infantile in comparison to how Nature is able to design, what is next? Do we press on, designing materials purposely separate from the environment and causing harm to it in disposal? Or can we seek a materials revolution in order to create materials of *good* rather than *less harm*?

TWINKLE TWINKLE LITTLE BOTTLE

On the surface, this seems like an impossible challenge to solve. Starting some 13.8 billion years ago, something special came from nothing in the form of hydrogen, helium, and trace amounts of lithium and beryllium, the four lightest elements in the universe. Over time, hydrogen and helium organized into gas clouds and then collapsed into stars, with the help of gravity, which led to the heating of its cores to produce fusion. Included in these stars, approximately 4.57 billion years ago, was the creation of our energy source, the sun; it is a relatively young star and provides us with our necessary hydrogen and helium atoms. Even in their collapse, older stars have become useful to us, sending their residuals to fall onto our planet, making our minerals. Can you imagine that you, your family, and your beverage containers originate from collapsing stars? Famous astrophysicist Neil deGrasse Tyson often notes that the elements that make up humans and other life forms (hydrogen, oxygen, and carbon) are in great abundance in the universe. Our beverage containers are made primarily from metalloids, defined as having both metal and nonmetal elements (glass), metals (aluminum), and organic material (plastic); how these different elements became our consumer products is the main story of this chapter.

In its origins, the earth was constantly bombarded by materials from the heavens that were necessary to stock nature's laboratory with the 118 elements that we know today exist on our planet. The sheer amount of inorganic material that has graced our planet is almost beyond comprehension; even after significant amounts of iron have been extracted and processed, the USGA has reported that world resources still exceed 800 billion tons (with annual world production less than 6 million tons a year).[5] Yet it wasn't until life formed on earth that the real miracle began, not just in the creation but also in the continual and advancing re-creation, which includes the reuse of millions of years of dead organisms that eventually became fossil fuels, and ultimately, our plastic drinking bottles!

Hydrogen makes up 75% of the ordinary matter of the universe. In turn, ordinary matter makes up 4.6% of the total universe (71.4% of the universe is dark energy and 24% is dark matter). Before its transformation, the early planet Earth had an atmosphere of hydrogen, methane (CH_4), and ammonia (NH_3) with no free oxygen. Largely through volcanic activity, the right balance of nitrogen and carbon dioxide for the development of life was provided to the atmosphere. From this, symbiotic relationships began to form, defined by these mysterious laws of nature that I will title the first material scientist, or Material Scientist 1.0. These laws of nature are our standard of good science, the formation of materials and life that is the reason for our existence, our modern society, and our beverage containers. An understanding of this evolutionary progression is not just a matter of physics, chemistry, and biology, but hopefully, the future for our development of materials that improve upon, rather than damaging, nature.

The greatest innovation for Material Scientist 1.0 was that of photosynthesis, the basis of all life on Earth. Photosynthesis is not a difficult concept to explain, in the use of sunlight to split water, yet its application is a bit of a miracle that cannot be replicated in scale and efficiency by modern science and industry. Stripping electrons from water to produce adenosine triphosphate (ATP) enables the energy source in cells to produce glucose, which is then used as an energy source for bacteria and animals. Through the radiation of sunlight, carbon dioxide, and water, simple sugars are formed via photosynthesis that originally led to the creation of life on Earth. Today, there is six times more energy created by photosynthesis in nature than there is in our industrial system model![6] It is in practicality the ultimate closed loop system that enables plant life to thrive and emit oxygen to animals, who then return the favor by emitting carbon dioxide for plants. These simple sugars that have driven evolution also drive our modern day economies through energy and materials.

The most important element to our organic world, of course, is carbon. This element is not only 18.5% of our own human bodies but is also the key to our most basic needs, such as food, and our modern necessities, such as energy. More so than any other element, carbon is able to bond freely in ways that not only lead to life, but humanity's synthetic materials as well. Lucky for us, carbon stored over millions of years became an almost infinite source of energy to power our modern world; the scope of this is almost impossible to comprehend: according to Yale geochemist Robert Berner, the entire living world of carbon life comprises 0.0004% of the total organic carbon, with the balance buried under the sediments over millions of years.[7] Unused millions of tons of carbon that were buried are now being harvested and incinerated for energy and materials in an alarmingly rapid fashion, impacting the ecological balance of our planet. This gift that triggered the Industrial Revolution and the post-Industrial Age is the primary feedstock for Frankenstein's bottle!

BEFORE MAN WAS GOD

Humanity's understanding of physics, chemistry, and biology is a case study of evolution itself; starting with fire, humans learned their most useful trick, creating the game changer to cook meat and plants, leading to larger brains, light and heat beyond the sun, and protection from other animals. After primitive humans moved into settlements and farming, gaining an understanding of harvesting organics roughly 12,000 years ago, humanity focused its attention on the use of inorganic metals and minerals for advancement. Of the 118 elements on the periodic table, 91 are metals, 8 are metalloids (in between metals and nonmetals, like silicon), and the remaining are nonmetals that are gases (such as hydrogen, helium, and oxygen) and solids (such as carbon and sulfur). During antiquity, those first days of material science, only seven metals were known to humanity, and only three (gold, silver, and copper) were found in their native form. Humanity's first practices in metallurgy were to extract these remaining four metals from their mineral compositions, given a limited knowledge and tools. Fire was the power behind primitive metallurgical techniques, being a limited energy source in comparison to the fossil fuels used in today's industrial world.

Fire has always been the foundation of advancement; according to Greek mythology, the god Zeus took fire away from mankind as a punishment until Prometheus disobeyed Zeus and returned it. As mankind's knowledge of fire grew in the Roman Era and Middle Ages, so did the advancement of mining techniques, processing, and management of energy. For example, people found that wood may only reach a temperature of 600° Fahrenheit, while carbonized wood, or charcoal, can reach temperatures of more than twice that, to almost 1,300° Fahrenheit! Charcoal processed from wood, peat bogs, and even surface coal was partially discovered during the Roman Empire, allowing for material innovations such as brickmaking, glassblowing, and more sophisticated metal production. In this era of discovery, humanity's mastery of materials reached an epoch that would forever separate us from other animals, but remained insufficient to achieve Industrial Era status; it would only be through the use of fossil fuels that creating true synthetic materials would be possible, which is the primary topic of this chapter.

High powered fossil fuels by themselves would be insufficient: an understanding of scientific methods would be required, leading to the Age of Enlightenment and a discarding of the mystical and secret traditions of the Middle Ages. In the year 1660, The Royal Society, today's oldest and most respected body of science, was founded upon the words *Nullius in Verba*, or "on the words of no one," building the case for science to be above societal institutions, such as the Church and State. Early developments in theories of atoms, molecules, and chemical reactions were forming; in the late 18th century, Antoine-Laurent de Lavoisier identified the critical elements of oxygen, hydrogen, and silicon, also finding that when matter changed its form, its mass would remain the same. In 1789, Lavoisier published a list of 33 elements, grouping them into gases, metals, nonmetals, and earths. Man was on his way to becoming Material Scientist 2.0; a sub-atomic understanding of science enabled him to create materials without divine intervention.

With knowledge and an almost inexhaustible, inexpensive energy source, the mechanized world of the Industrial Revolution was almost possible. Into the 18th and even the 19th century, however, energy and work were limiting factors to development since these were defined by human and animal power; simple mechanization driven by muscle; wind and water; and fire from biofuels such as wood, charcoal, and peat. A rise in population and the limitation of wood sources put iron and food production in jeopardy of not meeting growing needs. In 1798, economist Thomas Malthus argued that the population was growing geometrically and food arithmetically; therefore, it would not take long before famine would occur, creating a grave threat to mankind. Malthus was philosophically correct but practically wrong: in a decade or so, the fossil fuel energy source would transform the equation, leading to an explosion of growth in food and material production sufficient to keep up with the population growth (shown in Table 2.2). It is interesting to note that 75 years later in 1873, French professor Augustine Mouchot predicted the end of growth due to a lack of natural resources, namely the concept of *peak coal*; this questioning of the limits of natural resources continues today while technological innovations always seem to step in before calamity strikes. Let's hope we're up to the task in the 21st century!

Table 2.2   Changing Nature of Energy in England, as a Percentage

	Human	Animal	Firewood	Wind	Water	Coal
1561–1570	22.8%	32.4%	33.0%	0.3%	0.8%	10.6%
1600–1609	24.6%	27.5%	27.9%	0.5%	0.9%	18.6%
1650–1659	22.3%	23.7%	19.0%	0.8%	0.8%	33.4%
1700–1709	16.2%	19.4%	13.3%	0.8%	0.6%	49.7%
1750–1759	12.9%	14.6%	9.8%	1.2%	0.6%	61.0%
1800–1809	8.1%	6.6%	3.6%	2.4%	0.2%	79.0%
1850–1859	3.7%	2.7%	0.1%	1.3%	0.1%	92.0%

Source: E.A. Wrigley.[8]

PROMETHEUS UNBOUND

The pieces were falling into place: a world stocked full of valuable natural resources, an entry level understanding of the sub-atomic to unlock their properties, and a powerful and seemingly infinite energy source available for work to extract and manufacture at a massive scale. As is shown in Table 2.3, advances in manufacturing and transportation would only be possible through the doubling and tripling of energy potential; once this great energy source was harnessed and unleashed, it would become possible for modern society to reach almost God-like accomplishments, such as a daily use of 1 billion throwaway beverage containers.

Prometheus was almost ready to be unleashed. First, energy needed to be transformed to work in an efficient manner to power the industrial machines required to enable today's supply chain networks; mechanization was important to transport raw materials and finished goods as well as to automate manufacturing beyond previous capabilities. In his heralded 1969 book, *The Unbound Prometheus,* David Landes labeled "the Industrial Revolution as the final victory of humanity over the constraints of natural conditions" due to new technology and social values.[9] Other economic historians paralleled the use of energy as liberating humanity from the constraints of the natural (solar energy flow) as consistent with the French Revolution doing the same in society; as a result, there was the promise, or rather, expectation, that economics will lead us to unlimited growth. English economist Tony Wrigley illustrated the correlation between energy consumption and economic growth, corresponding to the English Industrial Revolution. Wrigley called this "breaking free from photosynthesis," which meant securing a source of energy that did not rely upon the annual process of photosynthesis, breaking free from the natural world.[10] Yet the irony did not escape

Table 2.3   Energy Potential of Fuels[11]

	Net (GJ per Ton)	Gross (GJ per Ton)
Straw	13.4	15.8
Wood	13.3	14.9
House Coal	28.7	30.2
Anthracite Coal	32.6	34.3
Crude Oil	43.4	45.7

Wrigley that this liberation from photosynthesis was only provided "by gaining access to the products of photosynthesis stockpiled over a geological time span."[12] The "mineral regime" would shift us away from the *organic economy,* and all of its constraints of sun and seasons, and the laws of nature. It would lead also to an almost infinite growth path of world energy consumption, far surpassing any alternative source then or even today.

The transition from Material Scientist 1.0 to Material Scientist 2.0 was now complete: the path of mankind's evolution was no longer dependent upon nature but rather purposely separate from it through an almost unlimited and powerful energy source in the form of hydrocarbon fossil fuels. Economist Nate Hagens calculated that this harnessing of energy is akin to each of us having 600 invisible slaves constantly at our beck and call to drive us to work and school, heat our food and cool our drinks, make our clothing and ship it from poor parts of the world in order to make it less expensive, and, allow us to drink from 3 one-time use beverage containers a day and throw them away.[13] Yet just as Viktor Frankenstein immediately understood at the sight of his creation, there was more to the design than understood.

BRITTLE BOTTLES AND MAGIC METALS

America's consumer and throwaway society is built primarily upon one thing: the cheap and abundant supply of oil for energy and mechanized work. In the words of FDR's Secretary of Interior John Ickes, "There is no doubt about our absolute and complete dependence upon oil. We have passed from the Stone Age, to Bronze, to Iron, to the Industrial Age, and now to an Age of Oil. Without oil, American civilization as we know it could not exist."[14] Without these fossil fuels of congealed ancient energy, our modern economy would not be possible, including our mass production consumer beverage culture! The transformation of materials, both biotic and abiotic, into beverage containers relies on the extreme heats possible through fossil fuels, the use of these fuels as a feedstock for our modern-day supply chain systems, and the high speed mechanization to form, fill, package, distribute, and dispose of containers.

A requirement for inexpensive, reliable, and sanitary portable containers to store, transport, and preserve food and drink began in the late 18th century; Napoleon's government offered a 12,000-franc prize challenge to the first person to invent a method of preserving food for the military.[15] Fifteen years later, Frenchman Nicholas Appert developed an approach to sterilize food through the use of glass bottles just before Louis Pasteur discovered that heat would kill bacteria in food. A year later, Englishman Peter Durrand patented the concept of using tin cans for preserving food, and then ten years after that, a tin-plated can. At the same time in the United States during the Civil War, glass was used for whiskey and mason jars for home canning, and tin-plated cans for Borden's condensed milk. Due to the cost of producing and distributing these materials at the time, these commodities were reused as a rule, leading to problems of supply chain cost and viability as well as sanitation and durability.

In the 20th century, things began to change dramatically: the first automated glass-bottle production line was in place with the Owens Automatic Bottle Machine in 1906. As a result, in 1899, there were 7.7 million bottles produced in the United States and in 1917, 24 million.[16] In the automated glass-production process, molten glass reaches 1,900–2,200° Fahrenheit, enabling it to reach the plastic stage, and a gob mixture is dropped into a mold for forming. Note that only through the energy density of fossil fuels is this process of plasticizing the materials possible. Today the ingredients of a glass bottle are plasticized to 80% silica, 10% calcium oxide, 10% lime, and trace amounts of other materials such as aluminum oxide, ferric oxide, barium oxide, sulfur trioxide, and magnesia. Before melting, around 15–20% of cullet (recycled glass) is added to the mix. The *glass transition* process occurs, taking these materials from a brittle to a molten or rubber-like state through very high temperatures; it is this same process that is used in the manufacturing of plastics. Glass is really not a solid or a liquid but rather an amorphous solid, which means it's somewhere in between the two.

In its initial use, glass bottles were only distributed as *returnables*, two-way bottles of thicker grade in order to withstand as many as 20–25 trips before being rendered unusable. Milkmen, and beer and soda delivery trucks would take the beverages to consumers, who would return the containers to the manufacturers in order to be reprocessed and then refilled. Environmentalists often question why returnable glass bottles are no longer used in the United States as they are in Europe, and the answer is that these heavier, brittle two-way bottles are no longer considered as efficient across the entire supply chain system in the United States. In Europe, there are shorter travel distances and a different beverage consumption culture than exists in the United States, where consumers demand lighter, more portable containers. Studies conducted in regard to the net sustainability of returnable glass bottles give mixed results: some studies find that two-way bottles are sustainable, given the opportunity to use the same bottle 25 times before remelting and processing. On the other side of the argument are the higher transportation costs for heavier materials and the significant amount of resources (namely energy and water) to sanitize the bottles prior to refilling. In nations such as Sweden, with a 92% collection rate of glass bottles, a supply chain model has been established with shorter travel distances and higher costs through regulation in order to make it viable. In contrast, the national supply chain system of the United States, which relies heavily on transportation, cannot cost justify two-way bottles from an economic standpoint, leading to a recycling method of one-way use, collection, and reprocessing through crushing and reforming.

The glass bottle industry often publicizes that its products are "100% recyclable," and made from organic materials, implying that it is up to the consumers to improve the woeful 34.1% recovery rate in the United States.[17] In 2008, the Glass Packaging Institute (GPI) established a goal of 50% reuse by 2013, and it missed the target with a 2014 result of 33.6%.[18] Explanations for why this target was not achieved are predictable: a lack of mandated bottle bill collection programs to increase recovery rates, and the use of commingled single-stream recycling (rather than separated and processed) that increased contamination and breakage. Despite claims of the glass bottle being "100% recyclable," the market equilibrium for glass bottles is 15–20%, meaning that anything above this percentage is deemed "supply chain inefficient." Therefore, while it is possible, and even optimal from a material science standpoint, to make a glass bottle from 100% recycled material, as the industry publicizes, it would be a prohibitively expensive activity, given existing recycling, industry and consumer practices in the United States, and even in Europe, where such practices are protected by regulation. Changing recycling and reuse practices and regulation to increase cleanly separated cullet for efficient reuse may trade one cost premium for another; mandating two-way use in the United States would do the same thing, putting higher cost pressures on the existing beverage supply chain. The design of the material and the consumer supply chain is a problem of the math not working, leading to a 33.6% reuse rate.

A 2009 study found that only 40% of glass collected from commingled, single stream curbside bins was reused for bottle-to-bottle or fiberglass manufacturing; 40% for lesser uses, such as roadbed filler; and 20% went straight to the landfill.[19] In the United States, there is a greater demand than supply for *clean cullet,* leading to such low bottle-to-bottle reuse rates as 15–20%. Other variables are factored into the equation, such as glass's material heaviness and brittle nature, leading to more damage and higher logistics costs, plus consumers preferring more lightweight, shatter-proof containers such as plastic and aluminum. In California, there are 800 million wine bottles generated annually in the state, with only a 40% recycling rate, largely due to the container being damaged in the process.[20] Clearly, this data shows that, despite the best efforts and interests in the United States and Europe to reuse glass bottles, the container was not designed to be reused at high percentages in today's mass-produced beverage supply chain systems. As a result, back-end recycling efforts are less than efficient.

The use of a tin can for food began in the late 19th century, and the American Can Company started considering packaging beer in tin cans a few years later. Unlike food, packaging carbonated beer in a tin can (actually tin-plated steel) was difficult, as the pasteurization process led to the cans' bursting; while food cans had to withstand 25–30 pounds per square inch (psi), beer required 80 psi. Another challenge was finding a can liner that prevented the metal from interfering with the taste of the beer. In 1935, a solution was provided, handling the pressure issue through a higher gauge can and a vinylite polymer lining to protect the taste of the product. In the first year, 200 million cans of beer were sold, which rose to one billion by 1941; it would take until 1960 for canned beer sales to surpass glass bottles.[21] However, with the onset of World War II, all metals were rationed, and beer production in cans was limited only to product sent to the troops. During the war, Swiss inventor Jakob Keller found a process to use pure aluminum as a replacement for tin, but it was a very expensive alternative and did not possess the strength properties required. After World War II, the United States government sold off its wartime aluminum plants to Alcoa's competitors in order to increase consumer use of aluminum. Moving from an economic model of material rationing to market inducement, and metallurgical improvements of alloying to strengthen the aluminum can, markets began to form; however, it wasn't until a major beer producer, the Coors Brewing Company, sought a market alternative to the tin can that the aluminum can became mainstream. Bill Coors predicted that aluminum was the answer to the growing quality and environmental problems associated with tin cans.

Unlike many of its competitors, Coors did not believe in heat-pasteurizing its product, holding that this killed some of the flavor derived from the brewing process. Bill Coors was also against the use of the tin can, believing its three-piece design hurt the flavor of the product while creating a waste problem, given the difficulty in its recycling. With the recent introduction of the one-way glass bottle in the United States and the growing environmental problem of littered tin cans across America, Coors considered the aluminum can both an environmental and economic solution. Going against the grain of most in the beer and soft drink industry at the time, Bill Coors believed it was his company's responsibility to ensure the recyclability of its containers. In 1959, Coors introduced the first aluminum can, replacing the three-piece metal can. In conjunction with its design, Coors established a recycling program for its cans, offering one penny a can for its return. Its "Cash for Cans" program led to an 85% recovery rate.[22] It was not long thereafter that Coke and Pepsi joined the fray in 1967, leading to a major shift in how Americans consumed their beverages.

Through Nature, aluminum is a nearly perfect material for the consumer products industry once it has been refined. It is the most abundant metal on the planet, composing 8% of the Earth's solid surface.[23] As a raw material, it does have some flaws, as it is not available in free form (most metals aren't), requiring intensive energy use to isolate it from its rock, bauxite. Prior to the use of fossil fuels to isolate aluminum from bauxite and to oxidize the material, pure aluminum was more valuable than gold due to its rarity[24]; in fact, the top of the Washington Monument is made from pure aluminum as testament to its pre-industrial rarity! Yet once this element is isolated from bauxite rock and oxidized through the energy of fossil fuel, it can be reused practically indefinitely without any degradation in specification and in a cost-effective manner; this makes it a superior material that can be considered designed for reuse, unlike glass or plastic. In comparison to iron, it is strong at a lighter weight and its oxide does not flake, so it is resistant to rusting. Aluminum is very strong at light weights, and is soft and durable, which has been the basis for its rapid growth in the late 20th and early 21st centuries. In 1978, *National Geographic* published an article titled, "Aluminum, the Magic Metal,"[25] presenting it as the material for the future. Not only for use in packaging, but also in the valuable sectors of building, automotive, aerospace, and the transportation industries, demand for this material is high; therefore, sending aluminum cans to a landfill is a poor economic practice. For its reuse across industries, aluminum is easy to alloy with other metals, such as magnesium, manganese (both important for a beverage can), copper, and zinc, increasing its flexibility in use. Today's aluminum can is super efficient when recollected, possessing an 88% yield from the process.[26]

Despite being the perfect material for reuse in the beverage-packaging, building, and aerospace industries, possessing a market demand of a 100% collection rate, aluminum cans in the United States are recycled at a 66.7% rate, well below its economic equilibrium point. When imported cans brought into the supply chain are factored out, the recycling rate falls even further, leading to a net domestic recycling rate closer to 53–55%.[27] On the supply side, the Aluminum Association has stated that the average reused content in an aluminum can is 68%,[28] considerably lower than what is economically viable. Aluminum-can maker Novelis has advertised its *Evercan* product to contain 90%+ recycled content, which is 4–5 times higher a recycled content than glass or plastic beverage containers. With demand for aluminum UBC from the building, automotive, and aerospace industries too, there is a shortage of UBC supply for reuse, yet close to 50% of all domestic U.S. cans are sent to landfills. How can it be the case that aluminum cans are collected, recycled, and reused at a much lower percentage than its economic equilibrium? Of course, one reason for this is due to the lower recycling rates in non-bottle bill U.S. states (39 states with a 23% recycling rate) versus the eleven bottle bill states (recycling rates of 70% and above). However, from my research, I have found that this does not describe the full extent of the problem that exists in mandated recycling programs for this valuable material. Through an overall recycling program strategy that focuses on all beverage containers, regardless of value, a lower than optimal percentage of valuable materials (aluminum) is collected while less valuable materials (glass and plastic) are collected above their market equilibrium. Ironically, while glass bottles are in higher demand than supply, they are collected at higher rates than market equilibrium in mandated systems due to the collection, supply chain, and manufacturing related issues, as noted above.

This is the recycling irony; aluminum appears to be the perfect beverage container material (except for hot drinks, of course), is in high demand, and has high (88%) reuse but is recycled at a rate well below its supply chain potential. If aluminum were the only beverage container material, or even if only aluminum and glass were available, this irony could likely be solved by bottle-recycling bills. There would be no recycling myth; a system could be put into place that balanced the reuse of bulky glass bottles and collected every aluminum can for remanufacturing. Instead, a new entrant, a disruptive innovation, would enter the scene and radically change the manner in which Americans consumed beverages. Next to arrive on the scene is Frankenstein's bottle.

FRANKENSTEIN'S BOTTLE

"Have you ever seen a polypropylene molecule?" a plastics expert asked author Susan Freinkel, author of *Plastic: A Toxic Love Story.* "It's one of the most beautiful things you've ever seen; it's like looking at a cathedral that goes on and on for miles."[29] Like the aluminum can, the plastic bottle originates from an abiotic feedstock, but unlike aluminum, this material is not a mineral. The feedstock for plastic is a fossil fuel, which isn't really energy but rather a good vehicle for storing it. Dead leaves, plankton, and other ancient organisms that have been buried for ten to hundreds of millions of years formed as hydrocarbons that pack a significant energy punch (as shown in Table 2.2). Legend has it that oil baron John D. Rockefeller was looking out the window and saw flames shooting from the smokestacks. He asked what was burning and was told it was an ethylene gas by-product from the refining process. He told the manager to "do something with it, I do not believe in waste," essentially leading to the development of polyethylene and other plastics. Thermoplastics, such as the polyethylene and polyethylene terephthalate (PET), can become moldable at a high temperature and then solidifies upon cooling. This would lead to a revolution in beverage packaging but also a supply chain system, including recycling, that acts separate from the laws of nature.

In the late 1960s, a DuPont engineer named Nathaniel Wyeth wondered why carbonated soft drinks could not be bottled in plastic as an alternative to glass. Wyeth understood that glass had to plasticize as it took form, and being a polymer scientist, he began to put together a case for change. Glass bottles not only presented a problem of higher supply chain freight costs as a function of the heavier material, it also was an inconvenience to the consumer. The material that Wyeth eventually used, a plastic called polyethylene terephthalate, or polyester, had been around since the 1940s as a synthetic fiber for clothing. Wyeth's innovation was not from a material standpoint but rather a process one: the blow molding process that would revolutionize how this plastic material could be used in the future. It led to the replacement of the glass bottle with a synthetic fiber that acted like glass but with superior overall qualities. From a front-end product development and marketing standpoint, the invention of the PET bottle was a radical market innovation.

Unlike the aluminum can, which was designed to be remanufactured as a metal, or the original returnable glass bottle that was to be reused up to 25 times, a PET plastic bottle was designed for only a single use. Conceptually, thermoplastics can be reheated and reshaped repeatedly, but food grade specifications make it much more difficult than in other industries. Wyeth's research was focused on developing a plastic composite where the polymer chains are designed to be more widely spaced to enable the flexibility required for a carbonated beverage. In 1973, Wyeth filed a patent for polyethylene terephthalate, or PET, to be used for plastic soda bottles. With this patent filed, plastic bottles began to be used in the beverage industry, and it is no coincidence that sales started to grow significantly, as did waste. Today, PET feedstock prices continue to fall in the United States, as a result of low natural gas prices and improved production processes.[30] As a result of improved supply and production processes, consumer preference, and subsidized recycling programs, this container material has become the overwhelmingly popular choice.

Technically, the PET bottle is made from a synthetic substance that originates from ethane, a chemical compound of C_2H_6 (two carbon and six hydrogen atoms). To arrive at these chemicals to make PET, ethane and xylene is extracted from the natural gas and cracked in order to build smaller unsaturated hydrocarbons. The first monomer is mono-ethylene glycol (MEG), in which the ethane chemical compound is unsaturated in order to have a double carbon bond without as much hydrogen included in order to become polyethylene. This process is important because carbon is the backbone of the polymer that will eventually be created in order to develop the strongest material possible at the lightest weight. Xylene, also from petroleum, is transformed into a crystalline solid called terephthalate acid ($C_{10}H_8O_4$). This monomer, which is 70% of the total weight of PET, is the most difficult one to create and therefore less able to replicate organically. When these two chemical compounds are heated together using a catalyst, PET is the end result that will later be spun into fibers or formed into plastic. (PET is a thermoplastic, so it can be reformed). This synthetic material becomes a super carbon structure—so strong, flexible, and inexpensive that it has become a ubiquitous part of our modern lives. The strength and lightweight combination of this plastic is both a blessing and a curse: the blessing in being able to protect a consumer beverage in such a lightweight and sanitary manner, and the curse in being so strong that it could last in our ecosystems for seemingly forever!

In many respects, the material science benefits of a PET bottle are similar to those of an aluminum can: thin flexible walls (to withstand the carbonation impact), with a strong polymer able to withstand enormous pressure. A single-serve PET bottle of 0.5 liters is strong enough to hold 50 times its weight in liquid.[31] Further developments have been made in design to increase lightweighting, with a 2-liter PET bottle that used to weigh 68 grams in 1980 now weighing around 42 grams, providing a huge advantage in transportation costs. PET also a good gas and fair moisture barrier, and is strong and impact resistant (unlike glass). The greatest difference between the aluminum can and plastic bottle is this: aluminum UBC is collected at a 50–60% rate, with the potential for a can to have 88% recycled content while PET plastic is collected at a 30% rate, with a bottle-to-bottle reuse rate of approximately 12–18%, meaning that it is overcollected, if anything.

As is the case in the glass bottle industry, the plastic bottle industry will correctly note that these thermoplastics can conceptually be made from 100% recyclable content, but in reality, there are concerns of food safety, supply chain efficiency, and material integrity in the higher reuse content. In comparison, aluminum can be reused over and over with no degradation in performance and specification, as well as being economical.[32] For example, if the recycled content of a PET plastic bottle exceeds 40–50%, a yellow colorant is visible in the bottle, impacting its color quality and form,[33] and there are more issues with food contamination the higher the reused content. The evidence suggests that the long, beautiful PET polymers were designed to resist nature and succeed within an industrial supply chain system, not to accompany nature and be reused; there are no known microbes in nature that can break it down effectively, nor back-end recycling and reprocessing centers from an economics standpoint. The industry must be honest about the fact that, in the material reuse of PET, the issue is not in recycling programs, but rather in the product's existing design.

Even in Europe where the collection, recycling, and reuse of packaging materials is subsidized by consumers and producers, only 30% of collected PET beverage containers are recycled into new bottles, despite a 52% recycling rate.[34] It is true that there is a higher reuse rate in Europe than in the United States due to more careful (and expensive) separation and reprocessing of these materials, but reuse rates in Europe appear to be capped due to the technical limitations of the material more so than by recycling program limitations. The National Association for PET Container Resources (NAPCOR) in the United States found a 31% loss in yield due to contamination from single-stream recycling, and 25% losses from refund-deposit programs. That leads to a low PET bottle utilization rate of 22%, not much different from reuse rates in the EU, where more careful collection and recycling procedures are undertaken. [35] In the United States, soft drink and water bottler giants Coke, Pepsi, and Nestlé have struggled with PET recycling and reuse programs, with only Pepsi achieving a consistent, yet low percentage of 10%, Nestlé at 8%, and Coke not maintaining a consistent percentage.[36] While these major companies often establish lofty goals, such as Coke's goal of 25% by 2015, or Nestlé's push for an industry goal of 60% by 2018, such targets do not appear to be realistic given the design of the material, regardless of what the corporate sustainability programs may envision.

Plastic bottles may be the largest category of the Frankenstein's bottle but is far from the only Material Scientist 2.0 design problem; there are also the take-out coffee cups, single-use coffee pods (e.g., Keurig K-cups), take out beverage cups from fast-food restaurants and convenience stores, and foil plastic sippy pouches. These less traditional portable beverage monstrosities are often not scrutinized as much as the dreaded plastic bottle despite these newer containers existing outside of scope of nearly every recycling and waste management program, thereby, leading to a growing market for waste. Perhaps the most visible beverage container that is conspicuously not recycled is the ubiquitous take-out coffeehouse cup, often portrayed by Starbucks, maybe the most respected consumer brand in the world. In 1984, entrepreneur Howard Schultz sought to revolutionize the coffee business through a plastic-lined paper cup in order to transform a boring morning-only drink into a high-end, all-day, hipster drink. Yet in order for this to be possible, a container would need to keep the heat of the drink on the inside for perfect consumption while ensuring the consumer's hand was protected on the outside.

Achieving a portable beverage container for a hot drink is much more difficult than a cool or room temperature drink. According to the Specialty Coffee Association of America, coffee should be around 180 degrees Fahrenheit; 25 years ago, a jury in New Mexico disagreed when a 79-year-old woman sued McDonalds for the third-degree burns she suffered when she accidentally spilled coffee on herself. The woman was hospitalized for eight days, undergoing skin grafting and two years of medical treatment. Her attorney found the coffee to be defective at a temperature of 180–190° Fahrenheit, and the jury awarded her $160,000 for treatment and $2.7 million in punitive damages.[37] In the end, a confidential settlement was approved after the judge reduced the award, but the message was clear to coffee retailers: protect the consumer from the heat. Enter the plastic-lined paper cup that prevents the liquid or the heat from soaking through while keeping the drink at the optimal temperature. To perform both feats, a Frankenstein container has been created that appears to be from a renewable source of paper/wood but is wrapped with plastic that prevents it from being recycled except in the most contrived situations. Oddly, this cup can be described as both abiotic and biotic, or probably more accurately, neither.

Today's plastic-lined paper cup begins with a paper specification that supports both stiffness and strength. Seeking to straddle the line between package protection, sanitation, integrity, and sustainability, Starbucks developed the cardboard sleeve to reduce the problem of double cupping; it also received permission from the Food and Drug Administration to use post-consumer content (10%) in its cup. Much like other consumer beverage companies, Starbucks is not shying away from the problem created with its take-out cup, facing an enormous challenge to maintain heat and food (contamination) safety, while at the same time having little control over the large percentage of containers that leave its store properties either as take out or drive through. In 2008, Starbucks made changes to migrate its plastic coating from polyethylene to polypropylene in an effort to reduce greenhouse gases, an incremental innovation for a company seeking a much greater splash (pun intended).[38] The company is also developing recycling/reuse centers to turn its cups into napkins and is seeking to increase recycling bins at all of its locations, both incremental innovations. It has ongoing efforts to find paper mills that can accept its cups and is holding "cup summits" with scientists and industry experts in order to achieve the disruptive innovation to safeguard its markets and the environment concurrently. In the interim, the company is painfully aware that 16 billion plastic-lined paper coffee cups are sent to landfills every year, with little to no recycling programs to support anything else. Yet the problem of safety remains the albatross around its neck in regard to cup recycling; in a 2012 incident, a North Carolina policeman sued Starbucks over a defective coffee cup, with the cup "folding in on itself," causing the lid to pop off and the coffee to spill on the plaintiff, leading to surgery to remove part of the man's intestine.[39]

One growing competitor to the coffeehouse drink is the single-serve coffee market, led by Keurig Green Mountain. Today, one in three

Americans has a pod style coffee machine at home, selling 9 billion of its little K-Cup pods on an annual basis—all of which are thrown away.[40] Like Starbucks and other consumer beverage companies, it outwardly acknowledges that these little pod throwaways are "not something we are proud of" and has plans for achieving full sustainability by 2020. The problem is the backward compatibility issue within its existing coffee maker models, where a differently designed pod would not work; once again, the issue of seeking to address the back-end waste after the front-end markets has been established arises.[41] Recently, Keurig announced a step toward recyclable pods in its K-Mug pods, with the main pod being made from a polypropylene material that can be recycled by the consumer separating it from the lid and filtering it for recycling; calling this "recyclable" is like stating your Powerball ticket may be a winner, which is an accurate, yet improbable statement.

The list goes on. Let's not forget about the multitude of fast-food restaurants and convenience store soda cups, and perhaps the kingpin of them all, the foil pouch kid drinks or sippy pouches. These multilaminate aluminum and plastic pouches make up over $500 million in sales annually in the United States and cannot be recycled anywhere in the world, despite possessing valuable aluminum.[42]

FRANKENSTEIN: THE MATERIAL DESIGN MATH PROBLEM

Today, our industrial supply chain system has achieved unprecedented success in material design to efficiently manufacture, fill, and ship over 1,000 packaged beverage containers per consumer per year for 318 million people in the United States. According to the Container Recycling Institute (CRI), the average American used around 250 beverage containers a year in 1970 and wasted almost zero, largely due to returnable bottles.[43] Today, consumption is well over four times this amount in use, and waste is over 50%, leading to a massive economic and environmental problem in how these used beverage containers are handled. And this is only in the United States. One-time beverage container use, particularly plastic PET bottles, is growing rapidly in the developing Asia Pacific region; in 2007, Western Europe and North America accounted for approximately 50% of the PET sales market, with Asia at 24%. In 2017, Asia will be 34% versus a combined Western Europe/U.S. rate of 39%.[44] With such high growth levels in less mature recycling markets, this waste problem of one-time use beverage containers that cannot be easily reused will only get worse.

Simply put, the creation of a material to be used once in a matter of moments should be considered a definition of bad science and bad math, given the influence of the consumer market and supply chain system on the overall design. No matter how many laws are enacted and mandated, bad material science and math must be replaced and cannot be mitigated; recycling programs can never solve the problem; they can only slow down the pace of economic and environmental misfortune. These programs should not be considered as good or bad, but rather insufficient to solve a much greater problem. Even worse, these bad practices prevent us from a stronger focus on better solutions for the future.

Chapter 3

The Throwaway Supply Chain

The junk merchant doesn't sell his product to the consumer; he sells the consumer to the product. He does not improve and simplify his merchandise. He degrades and simplifies the client.

—William Burroughs, American Author

WASTE AWAY TO PROSPERITY?

If Americans begin to recycle like other developed nations, such as those in the EU, can it avoid being such a wasteful society? In a 2011 New York poll, 87% of respondents say they recycle, but only 51% say that they do so every day, and the statistics suggest the reality to be even lower than this.[1] There is an obvious contradiction between our thoughts and behaviors, and the recycling enthusiasts advocate that closing this gap will address the problem. But really, there is much more to the problem of waste than recycling and the material science problems as noted in chapter two; over fifty years ago, social critic Vance Packard warned America about the shortcomings of a consumption-based economy: "The people of the United States are in a sense a nation on a tiger. They must learn to consume more and more, or they are warned, their magnificent economic machine may turn and devour them."[2] Recycling and consumption are often dragged into the conversation of waste and wastefulness, often blurring the terms to having the same contextual meaning. In this chapter, I will discuss how this has become an unintentional problem that has essentially taken the issue of waste off the hook.

Waste hasn't always been a major problem in the United States; once upon a time, American business growth was achieved through waste reduction, not the fostering of it. In the development of a sophisticated supply chain system after World War II, material design transformation (including plastics), waste management techniques, world-class manufacturing principles, and the proliferation of consumer marketing led to the transformation of the American culture from frugal to consumptive. In contrast to the frequent references from world bodies that "Americans consume too much and this is a sign of moral depravity," I find it to be a celebration of the remarkable supply chain transformations that have come together over the decades, albeit sometimes in excess and waste. Unfortunately, this culture of innovation has led to a culture of waste, and now it's time to use the skills of the former to eradicate the latter to balance economic growth and environmental sustainability in the future. In this chapter, I will present how and why America's 20th-century economic growth was driven by supply chain transformation and waste, including the tremendous growth of the consumer beverage industry. I believe that in order for us to understand how to protect the environment, we must first understand how our excessive supply chains can lead to collapse; then we can fix them. To do so, America must discard its throwaway supply chain, not its desire to consume.

WASTE IS A SIN

Prior to the 20th century, waste was not viewed as a personal choice, or even a foundation for economic growth, but rather a sin. The view's roots were deep within the American culture, and especially permeated Judeo-Christian ideology. Perhaps there is no greater example of how the American Dream and an abhorrence of waste combined than through the story of John D. Rockefeller, the oil tycoon of the late 19th/early 20th century. Rockefeller built the Standard Oil Empire after a modest upbringing, learning the lesson of frugality from his mother, who famously told him, "Willful waste makes woeful want." To Rockefeller, waste was not just a sin, but also a lost business opportunity to grow and serve others. Rockefeller's aversion to waste was legendary, and it paid off handsomely for him, profiting from the infantile U.S. oil industry that was unable to manage the inordinate amount of waste as a by-product of kerosene.

When Rockefeller entered the kerosene business, he was just one of thousands of entrepreneurs seeking to profit from this new and thriving business. In the late 19th century, kerosene became a suitable replacement for whale oil, which had been used for illumination for the rich;

moreover it was inexpensive enough to also replace candlelight for the masses. The refining process of crude oil required the distillation of the oil by heating, separating, and condensing the materials, keeping the kerosene and disposing of the rest. Many of the fly-by-night suppliers simply dumped the gasoline by-product into a river or creek, or just back into the ground, ignoring the danger of fires this caused. Ironically, the by-product that was disposed is what we know today as gasoline, a product that would eventually dwarf the kerosene market in sales—a story of how today's trash can become tomorrow's treasure! It would take a real visionary to transform such massive unwanted waste and environmental damage into a market opportunity of massive proportion, which is what John D. Rockefeller did. Hopefully, the future visionary who will solve America's waste problem won't be an unscrupulous monopolist as was Rockefeller in his later years.

Rockefeller was never considered an environmentalist; rather it was his religious upbringing that drove his aversion to waste. By refusing to dump gasoline, Rockefeller was encouraged to find alternatives for its use, such as the energy source for the distillation and refinement of kerosene. By using gasoline, he lowered costs, increased market share and distribution, and improved the environment by putting wasteful, inefficient suppliers out of business. Rockefeller's business plan was to first reduce waste and to improve the environment secondarily, with efficiency driving sustainability rather than vice versa. His plan was to market his kerosene (and eventually gasoline) product as high quality, reliable, and available to the masses at the lowest cost. By reducing waste, Rockefeller was able to make kerosene oil for lighting available to common people, enabling an improvement to their lives. Imagine how different the petrochemical market of the late 19th and early 20th century would have been if, in seeking to solve the problem, regulation had minimized its scope and reduced the incentive for entrepreneurs!

Rockefeller never received a medal of honor from any environmental group, but his business practices did much to stabilize the environment and lead to the larger economic market of the gasoline industry of the future. By using gasoline as a manufacturing fuel, Rockefeller gained an early advantage in the new market for the refinement and distribution of gasoline as a fuel in the automobile business, which became a burgeoning industry just a few years later. Eventually, the advent of electricity would revolutionize the lighting of homes and would greatly reduce America's need for kerosene oil, putting many suppliers out of business. Rockefeller's Standard Oil survived and showed tremendous growth, largely driven by its business practices built upon a reduction of waste and, later on, monopolistic practices. In 1883, Standard Oil's assets were less than $100 million with earnings less than $5 million, but in 1905, assets surpassed $350 million with earnings close to $100 million![3] Despite conventional wisdom, it is too simplistic for historians to pin Standard Oil's growth upon monopolistic business practices alone; while much of how Rockefeller consolidated market share could be attributed to this, his focus on efficiency was the driving factor in his initial success.

In the processing of a barrel of crude oil, Rockefeller was able to discover a vast array of products that others may have thought of as waste: petroleum jelly for medical companies, by-product waste to road construction companies, paraffin to candle makers and chewing gum companies, as well as materials for dyes, lubricants, paints, and acids. More uses led to greater efficiency, which reduced prices and pricing fluctuations, and increased market stabilization, leading to the growth of the petrochemical industry. Rockefeller was able to offer kerosene to the poor to illuminate their homes for 1 cent an hour. In his own words, this was a religious obligation, noting that, "God gave me my money. I believe the power to make money is a gift from God, to be developed and used to the best of our ability for the good of mankind."[4] Waste to Rockefeller was not only a sin but also a lost opportunity to serve others. Consider the contrast to today's throwaway society where waste, not its reduction, is the manner to serve markets.

Rockefeller may have been the best known industrialist to adopt this ideology, but he was not the only one. According to historian Samuel Hays, there was a religious overtone to the conservation movement based on the "gospel of the efficiency."[5] This ideological aversion to waste crisscrossed from government to business and the environment. In his famous book, *The Principles of Scientific Management* written in 1911, Frederick Taylor started by noting, "The conservation of our natural resources is only preliminary to the larger question of national efficiency."[6] In this tie-in between nature and industry, Taylor advocates that "the system should be first"[7] in reducing waste, much like what occurs in an ecosystem. President Roosevelt, Frederick Taylor, John D. Rockefeller, and Henry Ford were notable conservationists who advocated a balance and level of efficiency between nature and industry. "Waste is a sin" was the mantra that extended across boundaries of business, religion, government, and nature. Some of these industrialists certainly had their transgressions, but when it came to reducing waste, there was no separation between it and business growth; rather, it was one and the same.

CONSERVE OR PRESERVE?

How have we migrated from an ideology where waste was attacked and reduced in order to serve the common good to era of *growth through waste* with the use of environmentalism and recycling to clean up around the edges? It's not easy to say, but in the early days of the 20th century, a chasm began to form between those who wished to prioritize the economy over environmental concerns and vice versa. One of the first battlegrounds was the federal government's plan in 1913 to create a dam in Yosemite National Park to serve the growing city of San Francisco. Teddy Roosevelt was the president at the time, a man who has been admired for his conservationist approach to the environment and the elimination of waste. The conservation of nature was done as a matter of religious belief, scientific management, and general respect and reverence for the natural world. One of Roosevelt's leaders in this conservation movement was Gifford Pinchot, who later became the first leader of the U.S. Forest Service. His father, James Pinchot, was a successful businessman, primarily in the wallpaper and lumber industry. Regretting the damage done to the forests by his industry, the elder Pinchot endowed the Yale School of Forestry, the first of its kind in the country. He also sent his child Gifford to be schooled in Europe to learn proper forest management techniques. Drawing from his education, his father's lessons and religious beliefs, and his belief in scientific management, Gifford became one of the first reformists in the conservatism movement relating to forestry. Similar to Rockefeller and Frederick Taylor, he believed the best way to manage the environment was to reduce waste through proper use, not through protection. President Roosevelt put his faith in this approach as a matter of pragmatism. Even today, Pinchot is known as a reformer in the conservation movement.

For some, Pinchot was an example of an industrialist who learned from the mistakes of his predecessors regarding a necessary balance between the economy and environment. To others, like John Muir, he was a more subtle form of industrial development that would eventually wreak havoc on the Great Outdoors. Muir rebelled from his fundamentalist Christian upbringing, finding nature to be his temple that drove his faith and convictions. Yet it wasn't until he temporarily lost his eyesight as a sawyer through a saw slipping from his hand and striking his eye that he, figuratively, saw the light. Once he regained his eyesight, he vowed to "never see the world the same way again" and became America's first environmentalist.[8] Rather than trusting the laissez-faire business model of the early 20th century, Muir believed that nature was a temple of sacred ground that should be only visited, not intruded upon; he believed that industry would eventually compromise this sacred ground if allowed. He dedicated the remainder of his life to preserving as much nature as possible through the establishment of national parks; his efforts were successful through the eventual creation of Yosemite National Park in 1903. Closely related to these efforts, Muir cofounded the Sierra Club, one of the first grassroots environmental organizations in the United States.

In 1913, the battle lines were set in the Hetch Hetchy valley in Northern California between Gordon Pinchot and John Muir, two principled men focused on the environment from two different perspectives. Seven years earlier, the San Francisco earthquake and fire of 1906 made a statement that the city's water system was inadequate. To Gifford Pinchot, the preservation of nature should also include a sustainable purpose to society, while Muir believed that it should never be compromised. That year an agreement was reached under the condition that water and power from the dam be used only for public interest, but this promise was breached when the city began to sell power to PG&E. Muir's rally cry of "Dam Hetch Hetchy!" was the beginning of environmental activism that would define the Sierra Club for years. This little club that Muir founded in 1892 "to make the mountains glad" was unable to stand up to powerful industrial interests at the time, but today, as a $100-million organization, it is armed with sufficient money and clout to do so. Whether due to industrial promises not delivered, or a new environmentalist ideology, a policy wedge grew between saving the environment and growing the economy, rather than both focusing on reducing waste.

CONSUMING IS A VIRTUE

The first generation industrialists, for all of their faults, were able to solve some (not all) environmental problems as a matter of pragmatism, finding economic growth through a reduction of waste. After decades of waste reduction in industry and government programs, the "gospel of the efficiency" began to wane during the post-World War II era in America. Despite the storyline of confetti parades and happy reunited traditional families, there were real concerns that the ending of the war would lead to a slipping back into another Great Depression. Economist Paul Samuelson noted in 1943 the daunting prospect of the war ending and "ten million men reentering the labor market."[9] Another economist, Gunnar Myrdal predicted economic turbulence so severe that it would lead to violence.[10] Government programs, such as the GI Bill, took some pressure off the employment market, but the U.S. mega-manufacturing complex built during the war had the potential of falling back without a demand for the production of tanks, guns, and other wartime needs. In 1936, Hermann Göring noted that "guns will make us powerful, butter will only make us fat"; a decade later when he committed suicide during the Nuremberg trials, Americans were dreaming of butter after having been deprived for so long, and the government was beginning to think this was a good idea in order to avert an economic collapse. [11]

Culturally and religiously, Americans were thrifty by nature during this post-World War II period, but psychologically, after decades of wars and depression, they were ready to be unleashed—not just in consumption, but also in the liberation from the drudgeries of ordinary life, such as household chores and returning used materials, such as diapers and bottles. On August 1, 1955, a declaration of independence from drudgery was presented to America through a *Life* article titled, "The Throwaway Living."[12] The subtitle of this article told the story, noting that "disposable items cut down household chores." An illustration showed numerous throwaway products that seemed to be making the young family in the picture very happy. Advertised was a barbeque grill that was a meat cooker with an asbestos shell (yikes!), sure to save time in the kitchen. The article heralded the use of the disposable diaper, hypothesizing that this product alone would lead to an increase in the U.S. birth rate! Through the arm of an emerging marketing discipline, Americans in the mid-20th century were being permitted to consume as much as they wished and even to waste in the process. What was considered a sin a generation prior was now heralded as an emblem of liberty, the dawn of a new era! Gay Hawkins, an Australian researcher of consumer culture noted that in America "insufficient attention has been paid to how the development of retail packaging served as a material vehicle for the reorganization and transformation of food markets and consumption."[13] This was more than just packaging: it was a transformation of how Americans would live their lives.

According to Vance Packard, this was the moment when we jumped on the back of the tiger. Unleashing the consumer to buy toasters and cars can be seen as progress, but what happens to the economy when everyone has this stuff? For economic growth to continue, the equation would need to be reformed; consumers would have to keep buying, and the items they purchased would have to have a shorter lifespan. In some cases, they would become a single-use disposable item. The equation seemed simple, yet frightening. Consumers would grow in affluence and living standards with an increase in discretionary money, affordable consumer products in high supply, products intended to be replaced rather than fixed and reused, and the proliferation of packaging to rival the transformation of the goods themselves. The concept of conspicuous consumption seemed to be perfected in America, if not born there, as a marketing approach to ensure that manufacturing capacity, the engine of economic and labor growth, would always be matched to consumer purchasing. The basis for the argument is this: with socialism, you wait in bread lines, while with capitalism, bread lines wait on you! In the Cold War between the United States and the Soviet Union, capitalism and the democratic republic would become inextricably linked, with a consumer-based economy pushing for economic growth. In this ideological policy battle, the environment was not much more than an afterthought.

If consumption leads to waste that enables production that drives job stability and growth, is waste still a sin? While this question was asked at the onset, there was never an easy answer. In 1941, British author Dorothy Sayers wrote a piece titled, "The Other Six Deadly Sins" that questioned the Catholic Church's great focus on sexual morality, in contrast to the other sins, such as "the great curse of gluttony."[14] Perhaps it is a sin to want what one cannot afford, but what if doing so becomes a virtue in the industrial world to ensure continual economic growth? To Sayers, overproduction and overconsumption would inevitably lead to generating massive amounts of waste that destroyed values and substituted them with unsustainable actions. Consumers would lose a sense of personal worth and value to the community through their desire to purchase what they didn't need, which hurt the environment as well. In Sayers's *Creed of Chaos,* she noted, "A society in which consumption has to be artificially stimulated in order to keep production going is a society founded upon trash and waste, and such a society is a house built upon sand."[15]

Today, the average American is likely consuming twice what his counterpart was in 1960 when Packard wrote *The Waste Makers*, and that individual is consuming twice more than an individual prior to World

War II. Yet in some sectors, we are actually consuming less through economic advancement. Think of all of the everyday devices that have been replaced by your smartphone: camera, mobile phone, pay phone, Walkman, Dictaphone, computer, television, calculator, watch, alarm clock, navigational equipment, remotes, and more coming in the future. The point is, we must separate an argument that seeks to define and reduce consumption that is driven by innovation and focus on waste that is harmful to the environment. The throwing away of billions of plastic bottles into a landfill is harmful waste but can be good consumption if people are healthier by drinking more water, for example.

Yet the idea of *induced consumption* that started in the 1950s and 1960s was linked to waste from the start; back then, Paul Mazur, a Lehman Brothers banker noted that "we must shift America from a *needs* to a *desire* culture . . . people must be trained to desire, to want new things even before the old had been entirely consumed."[16] And yet it may not be all bad, as marketing expert Victor Lebow noted that America's "enormously productive economy demands that we make consumption our way of life."[17] But when does consumption lead to waste rather than productivity is the question, and it appears that America may have passed this point, requiring new ways of thinking about our existing supply chain systems. Today's way of thinking can be summarized by Kim Holmes, Director for the Society of Plastics Industry (SPI), who noted that "plastic has become a hallmark material of our throwaway society . . . requiring a supply chain holistic approach for solution."[18] Once the magnificent innovation of our modern supply chain systems are used to enable waste, it is time to reflect that we have likely passed into an era where the cost is greater than the benefit in our consumer-based economy!

PACKAGE FIRST, PRODUCT SECOND

In just a few short generations, Americans went from being waste prudes to waste gluttons. In Packard's book *The Waste Makers*, he tells the story of the American pioneer who cherished and guarded his bread pans and iron pots, given these supplies were hard to come by. Fast forward to the 1950s to a push from the aluminum industry to replace these pots and pans with single-use cooking containers that could be thrown away.[19] Prepared foods were becoming the rage, focused more on convenience than the quality of the product. Kraft was the first to offer individually wrapped slices of cheese, not to improve upon the quality of the product, but rather to expand distribution and lower market costs. Kraft's innovation was perfected by Standard Packaging in the early 1950s to vacuum pack the individual slices in order to increase the shelf life of the product. Marketing became a function of packaging and convenience. This led to a market growth revolution for food companies like Kraft and packaging companies like Standard Packaging, the company that provided disposable packaging in the food industry. In fact, Standard Packaging was able to triple their sales in four years, making disposable food trays that could be cooked, bags that could be boiled, and single-use utensils that didn't need to be washed, just disposed. The company's marketing mantra was that "it (the company) does not make anything that cannot be thrown away," creating a paradigm shift that it is better to waste and consume more than to spend one's time cooking and cleaning in a kitchen. The package became more important than the product, perhaps leading to lower quality, but greater convenience.

The packaging revolution was underway; an emerging restaurant named McDonalds was able to lower its costs by eliminating plates, glasses, and silverware and replacing it with disposables that enabled the customer to take the food out of the restaurant if so desired. On trains, bacteria infested glasses were replaced with one-time use Dixie Cups, a wax lined paper cup. Consumers were able to watch television shows while eating their TV dinners on TV trays, reducing the time that mom spent in the kitchen to both cook and clean. And of course, along came the single-use disposable beverage container that made refreshment possible any time, anywhere. More research and development focused on the packaging than the product, and all of this individualized disposable packaging did something that was unplanned: it led to a *reduction* in the cost of the product through a more efficient supply chain system!

This concept of *creative convenience*, to borrow a phrase from Charles Mortimer,[20] chairman of the board of General Foods, led to a new culture of how Americans would eat, drink, and live. Back in the 1950s and 1960s, when families remained a homogenous nucleus of a dad, mom, and two kids living at a less hectic pace, creative conveniences of carryout plates, plastic forks, and aluminum drink cans was a treat; today, it is a veritable way of life. The overscheduled lifestyle of a single parent or two working parents with kids in many scheduled activities, disposable packaging, especially beverage containers, is a ubiquitous part of our lives. Waste has become an invisible cultural necessity and no longer a religious aversion. Today, we are as far removed from the "gospel of the efficiency" as we can possibly be. The throwaway society is no longer just marketed: it is expected.

DISRUPTING THROWAWAY-ISM

Can recycling programs in themselves change the nature of our throwaway society? Of course, the answer is no; if poorly designed packaging materials are growing as one-way-use items, recycling programs become little more than waste management mitigation techniques that leads to landfilling and incineration rather than reuse. In forty years of recycling programs, the gains have been incremental, at best, in the face of mounting consumption and waste. Packaging recycling rates are flat to declining while personal consumption continues to rise, leading to more waste. Packaging waste permeates our supply chain system: aluminum cans that are valuable in a secondary reuse market (for packaging, aerospace, and building) are thrown into landfills while plastic and glass bottles are forced recycled at rates higher than their secondary market value justify. Today, there's a growing focus on bio-alternatives as the next great solution (as I will discuss in chapter five), yet I wonder why we cannot return to a policy of conservation that separates consumption from waste. Through innovation, consumption need not be obstructed, just waste; today, we're caught up in a model that views economic growth through consumption, and consumption through waste. We need a more logical approach moving forward.

Can we return to policies that promote innovation and efficiency that lead to economic growth and sustainability? To do so, we must replace the paradigm that economic growth can only occur through the creation of waste. Can market innovation solve the problem of America's throwaway society and packaging waste, akin to the problem of gasoline waste scattered across the landscape of the Midwest in the late 19th century? Yes, but it will take a modern innovator to solve the problem. As was shown in the past, environmental sustainability is best served when economic efficiency is the primary driver. With so much waste in our consumer markets, is this not a growth opportunity, as Rockefeller found in the late 19th century? Could it be Howard Schultz, and his mission to get rid of the Frankenstein Cup, or perhaps the future Elon Musk of packaging? Maybe an inducement prize concept, like the X Prize, would entice someone, anyone, to solve this problem. Who will be the one to take us off the tiger of waste that will end in our demise if unstopped?

Too many are distracted by confusing waste and consumption; economic growth should be encouraged so as long as it's not wasteful. Evolution has always advocated growth and emergence, not prevention and regulation. In a market-based economic setting, it is neither practical nor viable to solve what is essentially a waste problem through limiting personal choice and consumption, even if such choices (such as drinking too much soda) may not be optimal from any perspective. To balance economic growth and environmental sustainability we should optimize both rather than promoting one over the other. Even if we wanted to propose a more restrictive consumption model, as exists in Sweden, the American public would not allow for it, as was confirmed in numerous past elections.

The problem we face as Americans is not the consumption of millions of gallons of soda a day, or even the selling of a commodity like water that can be pulled from a faucet, but rather the more than 700 million packaging containers landfilled every day. Let's not trade economic growth for environmental sustainability as a method of moving backwards rather than forward. Chapter eight of this book presents solutions from a modern day framework of these legacy innovators, using rapidly accelerating technology that will require environmentalists to change their approach as well. In 2004, Michael Shellenberger and Ted Nordhaus wrote the provocative essay "The Death of Environmentalism," which defined environmentalism as just another special interest group. The authors called for a focus on innovation to solve problems rather than regulations and politics. According to Shellenberger and Nordhaus, the environmental leaders are "anything but stupid . . . but as a community, (they) suffer from a bad case of group think, starting with the shared assumptions about what we mean by 'the environment.'" The authors opine that, in the end, environmentalism "must die so that something new can live."[21] Much like the kerosene producers of the 19th century, innovation leads to disruption that breaks the special interests of the conventional state for a better path that can enable both the economy and the environment.

The solutions articulated in chapter eight can be generalized as follows: replace the outdated version of an economic growth model from the post-World War II era that seeks growth from waste, and provide incentive for innovators to solve the problem rather than mitigating it through legislation, such as mandatory recycling programs. Instead of castigating the super-efficient waste management conglomerates that whisk away tons of our packaging waste per person, enable innovation to achieve real *zero waste*, or even better, waste that is good for the environment, as exists in nature.

In 2013, Pope Francis spoke of a "culture of waste" that involves not only things, but people as well, developing a unique philosophy of "human ecology."[22] In the 19th century, Rockefeller held the perspective that the reduction of waste was for the betterment of consumers, to enable the poor to purchase his oil. While some believe the pope's words are only a metaphor, others believe it to be true: an economic model built on waste not only alienates us from our environment, but also from each other, which seems to be increasingly the case today. Can't we do better than throwing away these empty containers every day by the hundreds of millions?

Chapter 4

The Happy Cup Fallacy

We live in a disposable society. It's easier to throw things out than to fix them. We even give it a name—we call it recycling.
—Neil LaBute, American Film Director

CARTOONS AND CAUSES

On a hot Saturday afternoon in the summer of 2008, my wife and I took our two daughters to a Disney Pixar movie titled *Wall-E*. Expecting to get a few hours of relaxation and maybe even a little shut-eye while my four- and seven-year-olds were entertained, I actually got a little more than I bargained for. It may have been a kid's cartoon movie, but it had an adult theme: two robots whirling around a completely polluted Earth for roughly the first thirty minutes without any words other than repeating each other's names: "Wall-E" and "Eve." It was the year 2105, and the Buy n Large megacorporation evacuated the entire human race from Earth, sending them to live temporarily on a starship in order to clean the completely lifeless planet. With Earth evacuated, the megacorporation sent hundreds of Wall-E robots to clean the planet, but eventually, they gave up on the task as hopeless and pulled all but one of the robots. This last loyal Wall-E robot spends the next centuries drearily cleaning the planet and breaking the monotony by finding artifacts from the human civilization as collectibles. Wall-E's greatest discovery was a live organic seedling, the only sign of life on an otherwise dead planet. Shortly thereafter, another robot named Eve is sent to Earth in search of vegetation and is made aware of Wall-E's discovery. These two robots take the prized seedling to the contently obese humans, who resist the recommendation to return to Earth and start over. As the plot thickens, the two robots work together to persuade the humans that there is still hope, and the movie-goers leave with a "save the planet" message. Okay, let's all the save the planet by first disposing our massive plastic-lined paper popcorn buckets and plastic bottles and cups into the trash receptacles before leaving the theater!

Nowadays, it isn't just products and services that are marketed. Religion, politics, and yes, even environmental messages are communicated to us through a marketing medium. At the end of the 20th century, the late management guru Peter Drucker found the U.S. mega church to be the most important social phenomenon in the last thirty years, driven by marketing. We know that marketing has become a critical tool in politics, with campaigns more expensive than ever, requiring significant amounts of fundraising to win office.[1] Barack Obama's 2008 presidential campaign is viewed by many to be a success in its use of branding and social media. With regard to balancing the economy and environment, both consumer product companies and environmental advocacy groups have used marketing programs as well, leading to the *happy cup fallacy*, which will be addressed in this chapter.

INAUGURAL CAMPAIGN: THE CRYING INDIAN (OR ITALIAN)

The story begins in 1919 when a young army officer named Dwight D. Eisenhower took a journey on the Lincoln Highway, the first road across America that goes from New York City to San Francisco. Eisenhower's experience was less than comfortable, and a few decades later when he became the Supreme Commander of the Allied Forces of Europe, he experienced a German highway system that was far superior. When Eisenhower became president of the United States after World War II, one of his top priorities was to build a world-class interstate highway system for both military and commercial purposes. This world-class national transportation system brought great improvements in many areas, such as a nationwide supply chain network, but it also highlighted a growing problem of littering in America. I remember from my childhood watching a motorist throw something, often a can, out the window while driving down the highway. Back then, beverage packaging consumption was a mere pittance of what it is today but already a growing problem; in 1960 disposable packaging made up only 3% of the beverage market, but by 1966, this amount would rise to 12%.[2]

To address the problem of litter more than waste (a policy error), the Keep America Beautiful campaign was founded from a consortium of consumer beverage companies, nonprofit organizations, and government. The goal of Keep America Beautiful, or KAB, was simple: promote the concept of individual responsibility regarding the effect of litter on the environment. This puts the onus of recycling on the individual rather than the company, and the term "litterbug" was invented in the 1950s to describe the consumer's callousness toward the environment.[3] Around this same period, the environmentalists took to a different campaign, focusing on container bans and more stringent regulations on producers and consumers rather than public service announcements. Their efforts started off successfully: a ban on nonreturnable bottles was enacted in Vermont and Michigan, while refund-deposit programs began to sprout in other states. To these groups, personal accountability wasn't sufficient to solve the problem; government mandates were required to address the waste management problem.

According to the environmentalists, mandatory recycling program momentum was thwarted once industry adopted the use of marketing campaigns. The advertising firm Burson-Marsteller was hired to create an anti-litter campaign for KAB in time for Earth Day 1971, and boy did they hit the nail on the head. They released the Crying Indian campaign that became the definition of individual responsibility for recycling for anyone over 50 today. In this ad, a Native American played by Iron Eyes Cody is going through his day in the wild, and he keeps running into an abhorrent amount of litter and pollution of differing sorts—trash in the river, smokestacks from industry, and trash on the side of the road—which brings a tear to his eye. What better way to make the point of our trash problem and the consumer's responsibility than to portray a Native American amidst his ruined mystic landscape? We remember the catchphrase at the end of the ad: "People start pollution, people can stop it." What's magical about this ad is the guilt it leaves the viewer carrying: the Native American is aggrieved by the white man once again. Through this marketing campaign, it is the consumer, not the producer, who is on the hook for the problem of litter in a consumer-based society.

Even today, over forty years later, Keep America Beautiful is seen as the most believable of all the environmental groups, coming in first in a poll (36%), higher than the Nature Conservatory (29%), the Sierra Club (17%), Greenpeace (15%), and the Environmental Defense Fund (3%).[4] In the court of public opinion, it seems as if the key to an environmental message to the American public is simplicity and the least intrusive approach, such as planting trees and removing graffiti as opposed to changing one's lifestyle. To the environmentalist, the Keep America Beautiful campaign is the granddaddy of environmental greenwashing marketing programs: a faux concern regarding the environment while promoting an increase of production that is creating the pollution and trash in the first place. Antagonists considered the Crying Indian message to be a counterattack to legislated recycling programs, while protagonists suggest it as a proven method to reach the American public. One piece of trivia: the actor in the Crying Indian commercial wasn't even a Native American but rather was of Italian origin and was named Espera Oscar de Corti! Recently, KAB has implemented another successful program titled I Want to be Recycled, which anthropomorphizes a beverage container for effect.[5]

IT'S NOT EASY BEING GREEN, SOMETIMES

Marketing for organic food products and services is big business, estimated to grow to $211 billion by 2020.[6] Yet despite this massive growth spurt, green products compose only 1–5% of market share in many categories; for example, bioplastics are only 5% of the present plastics market.[7] Green marketing claims, such as "sustainable forestry" are becoming increasingly attractive to consumers, whether such claims are legitimate or not. Madison Avenue executive Jerry Mander calls such marketing practices "econpornography," highlighting the perspective that while it may be perfectly legal to do so, it may not be considered appropriate or moral to everyone.[8] By Earth Day 20 in 1990, 25% of all new product introductions were marketed as "green," "recyclable," "biodegradable," or "compostable," and today, practically all beverage containers presents this message on the label.[9] It seems as if virtually every market research study finds the environment and a company's reputation as critical to a consumer's buying habits; by 2009, 75% of the S&P 500 had a section on their respective websites about sustainability, and it's probably close to 100% today.[10] The AARP conducted a study and found that there are 40 million *green boomers*, which represents 50% of its base.[11] It is widely believed that consumers are paying greater attention to claims of being green, yet the jury is out as to whether they are willing to pay extra in order to be sustainable. In a study commissioned by Novelis, a leading aluminum manufacturer, about 70% of consumers noted that they would consider switching brands if one could demonstrate better environmental credentials,[12] but it does not address the real elephant in the room: are consumers interested in green products even when they cost more?

The term *greenwashing*, first used by ecologist Jay Westerveld, is the practice of disingenuously marketing a product and claiming to have policies that are environmentally friendly. Terra Choice, a Canadian environmental marketing company, found that 95% of products labeled as green in the United States and Canada are potentially misleading consumers, categorizing deceptive environmental claims into six distinct "sins of greenwashing."[13] Greenwashing, however, is a subjective term, and some see a conflict of interest from consumer watchdog groups, like Terra Choice, who profit from the creation of a "universal certification and labeling system." And there are numerous others focused on environmental labeling and classification: ecollabelling.org awards eco-labels, Green Seal, Energy Star programs in the United States, EcoLogo in Canada, and likely a hundred other organizations that claim to be the real McCoy when it comes to determining what's green or not. Environmental authenticity has become its own market sector through the creation of companies, councils, associations, programs, and public relations campaigns. All of this leads us to a battle of self-interests between those who wish to sell more products to consumers under the guise of green and those who wish to profit from classifying and assessing what is green and what is not!

It seems very difficult for the average consumer to cut through the claims, counterclaims, witch hunts, and definitions. What is the difference between biodegradable and compostable, or even products made from biomaterials and those that are biodegradable? In a hotel, should you rehang your towel after one use to help the environment, or are you just enabling a greenwashing campaign by doing so? Should a consumer products company create marketing campaigns that publicize its sustainability efforts in order to increase sales, or should it carry out better practices quietly, for ethical reasons alone? And if consumers really do care about environmental and sustainability reform, should they be willing to pay more for these products, or do they simply want to feel better through greenwashing practices? Given the intricacies of consumer behavior, market insights, and marketing, there are no easy answers to these questions; in beverage packaging, the happy cup fallacy is classified by some as a greenwashing exercise but by others as giving the consumers exactly what they want. Truthfully, it is difficult to argue with a narrative that consumers are more comfortable with less controversial environmental marketing messages, such as the Crying Indian campaign, than more radical programs that seek to modify how they consume in their daily lives, at least in the United States.

The "green police" seem to not only be on the lookout for companies but environmental groups as well: while partnering with the Clorox Company in a new line of green household cleaners in 2008, the Sierra Club was accused of a conflict of interest for receiving a percent of its sales. From one perspective, receiving $114 million for involvement and endorsement of these products was selling out on a Clorox greenwashing campaign, and to others, it was an example of collaboration between industry and an environmental advocacy group. Certainly, the Clorox Company did not make it easier for the Sierra Club to defend itself when it launched (in the same year) the marketing campaign "You don't have to be ridiculous to be green," putting the whole environmental approach under question. Was the Sierra Club a sellout, or were they seeking to partner with industry in order to achieve a balance between the economy and the environment? Being green often depends upon one's perspective.

Virtually all true eco-friendly product offerings come at a higher price, often beyond that of mainstream consumers; according to studies, there is a 16–100% sustainability tax associated with green products in comparison to standard product offerings.[14] Studies have shown that a typical Walmart shopper will spend as little as $65 a week on consumables, leaving little discretionary money to spare for higher priced items. Understanding the sensitivity of green pricing, Walmart relaunched its brand image in 2005 toward being green but in a different manner: by offering the moniker of "lowest prices, always" through a "waste is a sin" mantra, focused on waste reduction as its gift to both the environment and consumer. By reducing waste in its supply chain, lowering energy consumption and CO_2, and even offering some green products, Walmart's sustainability program is more of a throwback, offering the consumer value while helping the environment. In 2013, Walmart reached its target of a 5% reduction of packaging in its stores, which has added up to $250 million annually to its bottom line through recycling of corrugated board, bottles, cans, and more, and then selling it to recyclers.[15] Walmart's critics find this message to be disingenuous, citing concerns over low wages, a higher carbon footprint for consumers to travel to a Walmart supercenter rather than a corner store, and an obligation to lead real change, given its market position. While one can never make everyone happy, Walmart's green image seems to be a successful campaign with its consumer base.

Environmental advocacy groups must also consider their green marketing campaign in order to reach its target market. Take, for instance, the Container Recycling Institute (CRI), an association based on raising awareness of packaging waste in the United States. What I appreciate about CRI is its content-rich approach to addressing the problem of packaging waste, offering vital statistics and position papers on how to address the problem. While other environmental advocacy groups take a glitzy or even risqué approach to marketing its message, the CRI is like Wall-E the robot who performs its general awareness and campaign activities without great fanfare; contrast this to PETA's State of the Union Address campaign where a young female model undresses while lecturing the viewers on the animal cruelty of restaurant chains and circuses. CRI's website is full of data regarding the magnitude of the waste problem that is happening, presenting an articulate storyline for how our waste management systems whisk away our garbage before it hits our consciousness. Yet it's safe to say that most Americans are patently disinterested in statistics about packaging waste while millions are willing to partially listen to a message about animal cruelty for a trade-off of some eye candy. Shouldn't there be an outcry regarding these messages as well? To truly solve the problem, there needs to be an actively interested and informed public, regardless of marketing messages. To steal the words from Lorax from the great Dr. Seuss book from 1972, "Unless someone like you cares a whole awful lot, nothing is going to get better. It is not."

Greenmailing is pervasive in academia as well, creating a difficult environment for those who wish to challenge conventional wisdom. Upon completing my Ph.D. in 2010, I officially joined the academic ranks as a researcher and began to challenge the notion of sustainable supply chains and recycling programs that seemed to offer more from a marketing perspective than tangible, sustainable results. Upon studying the extant literature on the topic, I found an overwhelming display of papers that advocated the benefits of sustainable supply chains without clear articulation of what that really means from both an environmental and economic perspective. Believing this to be an opportunity to fill a gap, I began to submit papers to journals challenging this notion, not because I don't subscribe to environmentalism and sustainability, but rather due to a lack of clarity of what this means in industry. To my surprise, my submissions were often rejected for being "outside the scope of the journal," which is a nice way of stating that my agenda was not necessarily in agreement with the positions of the editors. I found an echo chamber amongst professors and researchers, who must publish papers in order to increase their "impact factor," a prerequisite for tenured positions and grant money related to green. Analogous to consumer product companies achieving market growth through sustainability or consumer insight groups raising revenue through eco-classification systems, academics were marketing themselves as a green product. Ironically, I find Europe to be a more accommodating academic model of dissent despite its stronger environmentalist brand. Upon presenting a disruptive innovation idea (solution 3 in chapter 8) to a well-established Norwegian professor, he scoffed, calling it a "gimmick necessary for an American recycling program to achieve European results," yet he was willing to understand its implications. I will discuss this tunnel vision in Europe in chapter six in regard to recycling programs.

THE *HAPPY CUP* OR THE *DAMN CUP*?

So far in this book, I have presented this recycling myth as a complex set of interwoven relationships between consumer product companies, consumers, environmental advocates, marketers, economists, material scientists, academicians, and even waste management companies. Rather than a conspiracy, it has become a market equilibrium where economic growth is driven through waste and others profit on the back of sustainability. In this storyline, the waste management industry has a difficult balancing act of processing and whisking away millions of tons of waste but doing so in a sustainable manner in order to be viewed as green, even though much of what is processed has little-to-no secondary market value. In a recent annual report, the Waste Management CEO wrote, "We're focusing on the hard choices confronting our customers, our company, and others who want to get to zero waste. . . . [T]o be sustainable over time, our operations must make economic sense."[16] In 2014, the company invested "virtually nothing" in its recycling operations, which accounted for only 17% of its revenues in 2013.[17] With most of its revenues coming from processing waste, not recycling materials, and even higher profits from the former and not the latter, it is logical for them to invest more in waste and less in recycling. Yet to acquire an image of sustainability, Waste Management announced a marketing campaign called Think Green that conveys an image of its landfills on or next to rolling hills of green, producing renewable energy, lush forests, solar generation plants, and wildlife preserves. Whether this is a greenwashing campaign or an appropriate balancing act is left to the discretion of the viewer, although it seems as if consumers find it to be exactly what they want!

Peter Senge's term, the *happy cup fallacy*, places producers and consumers in a symbiotic relationship where disposable beverage consumption is encouraged through a perception that these containers will be recycled, reused, and/or biodegraded when in fact, most will not. Whether it is due to a belief that recycling programs can work or it is just greenwashing, in the end, the consumers are rid of their guilt about using something for thirty minutes and then discarding it. Consumers are marketed to consume and then to recycle, but after that, nobody really knows what will happen. Perhaps it is marketing schizophrenia, but it works: the consumer is liberated to consume more so long as the container is placed into the little recycling bin afterward. Yet in the end, it doesn't really matter whether the container ends as trash or recycling.

There is no consumer product that better embodies the happy cup fallacy than the bottled water industry. In the Middle Ages, water was unsafe to drink, and, partially as a result, fermented beverages were preferred, leading to what must have been an interesting era, I would think! Eventually, with the advent of chlorinated municipal drinking water, public sanitation systems in the developed world would offer a clean, reliable water source to the general population. In the 1970s, there were growing pollution concerns in the United States, leading Perrier to launch a marketing campaign to a niche market for its purer drinking water, offered in glass, family-sized containers. It wouldn't be until the PET plastic bottle became a market device that this product would begin to take off, though, as is shown in Table 4.1. It was the portable, convenient bottle that led to the marketing success we know today as the throwaway water bottle.

Despite concerns brought forth by environmentalists regarding the damage these market devices cause, bottled water continues its upward

Table 4.1   The Escalation of Marketing Genius of the Plastic Bottle

Year	Bottled Water Sales (U.S.) (in billion units)
1996	2.8
1998	4.6
2000	8.4
2002	14.7
2004	23.6
2006	35.5
2008	39.9
2010	42.6

Source: Container Recycling Institute.

growth, with a compounded annual growth rate of 9.5% between 1976 and 2013, far outpacing every other beverage.[18] Despite over 50% of all bottled water originating from a municipal tap water source, with most of these municipal operations having higher regulatory requirements than bottled water, most consumers believe the commercial water is purer than the public water. This largely appears to be due to the marketing of bottled water as having a complex filtering process that disposes of its impurities. In this process, most bottled water companies have a three-step purification method of carbon filtration to remove chlorines, chemicals, and large particulates; ozonation that eliminates microbes, and reverse osmosis that removes metals and other contaminants. However, studies have raised some questions regarding how pure this water remains after packaging; a 2008 investigation by the Environmental Working Group found that some bottled water is inadvertently contaminated by industrial chemicals in the process.[19] From a marketing standpoint, the bottled water industry has conquered the debate, portraying a storyline of purified water that is portable enough for everyone to become healthier by drinking more water.

While virtually every bottled water company has marketed its commitment to reduce its use of packaging material, or to make the packaging more recyclable, it still faces a reality of being less sustainable than water from a faucet, drinking fountain, or other distribution points from our public water system. To combat this perception, bottled water companies have established marketing campaigns in order to grow its beverage market share from 14.4% in 2009 to 17.8% in 2014, with $12 billion in annual revenues in the United States.[20] The world's largest bottled water provider, Nestlé, has made the claim that its bottled water "is the most environmentally responsible consumer product in the world," far from an apologetic statement. Launching a new design tied to Earth Day, Nestlé's Poland Springs introduced an eco-shaped bottle designed to reduce its environmental footprint. This bottle is considered an eco-bottle due to its 30% reduction in plastic use, which is a good thing, or perhaps less of a bad thing, since most PET water bottles are trashed rather than recycled. Another sustainable marketing campaign for bottled water is for Fiji Water, the largest imported brand in the United States. Fiji claimed to be the world's "first carbon negative product" through taking more carbon from the air than adding to it. In a class action lawsuit brought forth against Fiji, it was claimed that its use of "forward crediting" of planting trees on the island (taking credit for what could happen in the future) is not a legitimate method for measuring a carbon footprint[21]—especially for a product that is shipped at least 5,539 miles from a tiny island to the United States. Further claims have found that while Fiji water produces more than a million bottles a day, more than half of its citizens do not have reliable drinking water.[22] Despite its questionable claims of sustainability, Fiji Water is the environmentally chic beverage for notable celebrities, such as President Obama and Paris Hilton, amongst others.

Coca Cola's Dasani brand has created its own use of green marketing through the introduction of the *plant bottle,* which will be discussed in greater detail in the next chapter. Making its debut at the Copenhagen Climate Change Summit in 2009, the plant bottle uses a plant-based feedstock rather than petroleum for part of the bottle's material. While the percentage of plant-based material varies from market to market, the maximum organic material in a PET designed bottle was approximately one third, given an inability to find an approach to replicate the terephthalic acid (TPA) portion of PET through organic means.[23] However, just recently, Coca Cola has announced a bottle produced from 100% sugar cane feedstock, completing its objective to achieve a 100% renewable (but not recyclable) bottle. Other claims exist as well: in the airport recently, I found a bottled water brand (Artic Sol) that noted its water bottle is made "from recycled bottles*," with the asterisk noting "a 10% minimum recycled content in each bottle," meaning it is greater than 10% and likely less than 15%. Practically every bottled water company has some form of green claims, which is understandable given the marketing challenge of selling a municipal commodity for $1–$2 a bottle, or even higher.

Bottled water is a market marvel, and so is the concept of Starbucks, a company that has transformed a rather boring morning product of coffee into an all-day customizable beverage. The brand is more than just a drink; rather, it is built around the "Starbucks Experience" that has a different meaning for those who chat or work using Wi-Fi in the restaurant, soccer moms using the drive-through, and even the younger generation. My teenage kids love Starbucks, as this brand speaks to them even more so than Apple, Google, or really, any other. As is the case in the waste management sector and bottled water industry, the challenge for Starbucks is the balancing act of offering an experience without burning the customer, and showing sincerity toward environmental stewardship. In a 2012 survey, 15% of millennials stated that they have made no effort to help the environment, a much higher percentage than the prior two generations.[24] While the younger generation may admit to making little effort, as I see through my kids and their friends, most of them will profess their concern for the environment, even if it doesn't show up in their actions. With this obvious cognitive dissonance amongst its consumers, how can a leading company like Starbucks stay focused on replacing the *damn cup*, as it has been unceremoniously called in company meetings?[25]

The coffee beverage industry throws 16 billion cups into landfills on an annual basis, which breaks down to 44 million every day, over 2 million every hour; of these drink cups, 4 billion are attributed to Starbucks, with 3 billion being paper cups and 1 billion plastic.[26] In 1984, Starbucks introduced the paper cup when it was just a tiny, regional seven-store operation. Today, the company is a massive $26 billion operation, with a 36.7% U.S. market share.[27] In 1984, Starbucks introduced the plastic-lined paper cup, which contributed to its ability to transition from a small operation to a large coffee chain. Later, it developed the cardboard sleeve to reduce double cupping and was the first hot beverage company to receive approval from the Food and Drug Administration to use post-consumer content (10%) in its cup; regardless of environmental concerns, food safety and sanitation is considered a greater concern in the reuse of post-consumer content. As a company, Starbucks rarely does any advertising or marketing for a good reason: its cup is everywhere, and no additional branding is required! According to Jim Hanna, Starbucks Sustainability Director, "our cup is our icon, our billboard, our ethos for the company. Customers have this great experience of interacting with store partners and the beverage. Then, when they're finished, they say, 'Now what do I do with my cup?'"[28] With 80% of Starbucks purchases leaving the store in a disposable one-time use cup, translating into over 3 billion thrown away in a year, the *happy cup* is the *damn cup*, embodying its brand image as well as its environmental problem. The company has held various "Cup Summits," even inviting its competitors, in hope of solving this problem; Howard Schultz has already missed his well-intended pledge of a 100% recyclable cup by 2012, and he appears to be bothered by this. So—it bears repeating— the billboard of the Starbucks cup is apparently as much of a *happy cup* as a *damn cup*.

In this chapter, my goal has been to articulate that the realities of consumer markets and environmental stewardship programs are not as cut and dry as many would like them to be. There is are very fine lines between a greenwashing campaign and well-intended sustainability programs, especially given that consumer research clearly shows that Americans talk more than act when it comes to recycling commitments.

Balancing economic and environmental objectives may have no more difficult a challenge than in the beverage industry, given its packaging and marketing of life's most basic commodity into a plastic bottle, the making of an ordinary morning beverage into a cultural experience, and the social nature of beer and wine more so than being just beverages. According to University of Indiana anthropologist Richard Wilk, buying a bottle of water is "buying choice, buying freedom, because that's the only thing that can explain why you would pay money for a bottle of something you can otherwise get for free."[29] What other industry has to use this level of marketing to grow its revenue? Those who cast judgment upon the beverage manufacturer need to understand these delicate challenges, and those in the industry must look toward the future of possibilities. A paradigm shift in how products are offered, packaged, and distributed must take place, one in which changes to the design of the container and supply chain enable real sustainability and market growth. When this occurs, we'll be able to change the title from the *happy cup fallacy* to the *happy cup reality*.

Chapter 5

Bottles Grown from the Soil

MANIC ORGANICS

If consumers are willing to pay over $2 for a bottle of water that can be pulled from the tap at 0.3 cents a gallon, why not have a 16–100% sustainability tax for green products? In 2013, organic food sales surpassed $35 billion in the United States yet are only 5% of the total food market per the Organic Trade Association.[1] Despite its fast-paced growth, many of the buyers of organic food and drink may not understand the value proposition in doing so; in a 2013 study conducted in Germany, most of the subjects did not understand what they were purchasing when they bought organic, which did not stop them from buying it. According to one of the researchers, Jo-Ellen Pozner, "People want something healthier, so they purchase organic. But they do not know what the term means, which leads to a watered down definition of the term."[2] In another study conducted by Alan Dangour that was published in the *American Journal of Clinical Nutrition*, 50 years of scientific evidence showed that organic food is no healthier than conventionally grown foods, yet during this period, the organic food market has grown exponentially.[3]

Regardless of the debate, the organic food market will continue to grow in the United States and elsewhere, as will organic clothing, household cleaners, and even packaging materials. From a marketing perspective, this makes sense; consumers are seeking to break from a world of mass-produced, artificial products and return to nature, even though these purchases may be more image than reality. Organic meats and produce are the most obvious choices, being free from pesticides and other harmful agents, such as growth hormones that are harmful to us and our environment. Organic cleaners, clothing, and packaging materials may be less obvious in regard to value at a higher price, but a growing number of consumers see these products as valuable, both from a good faith gesture to the environment and perhaps to their own well-being. These marketing programs and consumer behaviors lead to the question: Are today's organic bottles really better for the environment, or are they more of a marketing strategy, such as the happy cup fallacy described in the last chapter?

Prior to the "Age of Oil," as John Ickes famously called our present era, natural materials were used exclusively, including metals, stones, ceramics, and natural polymers, such as horns, bones, and wood. Leonardo da Vinci developed materials from various vegetables and animal parts, and for centuries, organic materials were known for their material properties. Polymers from nature, such as the cell walls of a plant, are simply long strands of monomer molecules repeated over and again in a flexible and strong structure. Synthetic polymers, often from fossil fuels, follow this same design but can be produced more abundantly, cheaply, and in more variety than in nature, creating more amenable polymers of greater utility. Bakelite was the first synthetic plastic to be developed, soon followed by polystyrene in 1929, polyester in 1930, polythene in 1933, nylon in 1935, and our plastic bottle material, polyethylene terephthalate (PET) in 1941, which was first used for bottles in 1973. At the end of World War II, the manufacturing capacity of petrochemicals shifted from wartime to consumer use, in fuel for automobiles and materials for consumer goods. This was no problem; not only was there significant pent-up consumer demand for new products, as was mentioned in chapter two, but the fossil fuel-based plastics were cheap in extraction, manufacturing, and distribution yet achieved a wide range of material properties, ease of use, and sanitation. Petrochemicals were used in the growing of food as well; the large supply of ammonium nitrate that was used for the production of bombs became a fertilizer. Indian food activist Vandana Shiva notes that, "we're still eating the leftovers from World War II." This Age of Oil would bring about an industrial society of synthetics in all aspects of our lives, separating nature from our sanitary and convenient world. Plastics, unlike natural materials, could be cheaply manufactured into almost anything, as is evident from the following exercise: look around where you are sitting right now, and count the number of items made from plastic (if you can count that high)!

GROWING BOTTLES

Ironically, the same unbridled enthusiasm that happened after World War II with the onset of synthetics appears to be happening today in reverse; instead of separating ourselves from nature, this mania for organics is a reconnection to photosynthesis, so to speak. The growing of bottles is a reaction to the weirdness of plastic based on the belief, albeit a fallacious one, that using an organic feedstock polymer rather than a fossil fuel will lead to a plastic water bottle that is recyclable and reusable, as if it were a plant itself. In the 1970s and 1980s, there was an opposite mind-set: the desire to use fossil fuel feedstock with optimal properties to manufacture a beverage container that was nature-proof and to create composites of virtually unlimited specifications of strength, weight, and flexibility, at a low cost. The Du Pont engineers who created this synthetic polymer did so with the goal to be virtually indestructible by nature, with no organisms in nature interfering, thus enabling food safety and supply chain transformation. Consumerism was enabled through improvements in manufacturing and an almost unlimited and flexible feedstock of fossil fuels. Today, there is movement toward an organic feedstock, using a similar manufacturing process to plasticize the material to create something that commences as industrially nature-proof but then becomes naturally symbiotic after use. We want the best of both worlds even though these worlds have been diametrically opposed in our current supply chain model. What's the miracle that can make this happen?

Today, there is a perception in the consumer beverage industry that the replacement of fossil fuels with organic materials is the answer to the problem of low reuse rates of containers. Fossil fuel, by its very definition, is an organic substance, but one of an ancient material; therefore, it is not renewable. Fossil fuel is made from plants and other dead organisms buried in the soil, compressed over millions of years to effectively become congealed sunlight, or energy. Therefore, while the fossil fuel that is used to make a plastic bottle was at one time an organic (carbon-based) material, like today's organic feedstocks of corn and sugar cane, the former is not renewable and leads to higher CO_2 emissions. Yet neither end product is highly recyclable and reusable as will be discussed in this chapter; the bottle grown from the soil is more of an incremental innovation with the perception of improvement being greater than the reality. As I noted in chapter two, Man as Material Scientist 2.0 has yet to crack the design of the beverage container to exist in symbiosis with nature, regardless of whether the material is organic; this is not just a material question but a process manufacturing and supply chain system question as well.

In theory, a plant-based container should have lower CO_2 emissions than a fossil fuel-based feedstock, given that the former is renewable while the latter is not. Photosynthesis in the growing process should offset the CO_2 that is generated in the development of a bottle; therefore it is assumed to be carbon neutral at worst. But in our modern agriculture system, we must account for the total use of fossil fuels in the supply chain process, including the use of pesticides and fertilizers as well as the fuel costs associated with the logistics of our agricultural networks. According to food expert Michael Pollan, it takes 10 calories of petroleum to grow one calorie of food; therefore, if that one calorie of food, such as corn, is used to grow a plastic bottle, is the plant bottle still carbon neutral?[4] At present, there are insufficient research studies to determine if today's plant bottle is more or less sustainable than the conventional PET fossil fuel-based container. According to the journal *Nature*, one third of all greenhouse gases emitted are due to agriculture, which offsets the photosynthetic growing process to be calculated.[5] This is only from a CO_2 standpoint; from a recycling/biodegradability performance standpoint, the renewable and nonrenewable feedstocks perform essentially the same after they are plasticized into long polymers of synthetic material. In fact, given the inefficiencies of the current state recycling systems, adding more variation through these organic inorganic bottles actually reduces recycling rates and increases its costs. From an overall supply chain perspective, an organic feedstock, such as corn or sugar cane may not be more sustainable than a petroleum based container when all factors are considered, including water use; at present, I would classify the research as inconclusive regarding which one is better or worse. The point here is that we shouldn't assume that a beverage container is more sustainable simply because it originates from a renewable feedstock.

Beyond the material science and supply chain factors involved, we must also take into account the economic impact of commodity markets on the design of our beverage containers. When the price of a barrel of oil flew past $100, there was a strategic push to wean ourselves from fossil fuels and the plastic production that accounts for 5–10% of its use. A few years ago, the International Energy Agency (IEA) prognosticated that a barrel of oil will reach $200 by 2030, which was seen as a threat to the national economies.[6] Major efforts were undertaken to find fossil fuel alternatives, which led to the push for the use of biomass in the packaging sector. In 2015, oil fell to as little as $43.39 a barrel in March, almost three times lower than the 2014 peak of $107.95; with this fallen price, and the proliferation of natural gas and shale oil in the United States, among other world events, much wind has been taken from the sails of the green packaging movement. Natural gas prices have traditionally followed the path of crude oil prices but, in 2009, took a sharp departure given increased U.S. domestic production. Now with both crude oil and natural gas prices falling precipitously, it is clear that the decreasing interest in green packaging may be tied as much or more to petroleum economics as environmental sustainability. The question is whether Americans care much about recycling programs and alternative bio-bottles when fossil fuel prices are so low.

While it may be true that America's growing energy role in the world may reduce the incentive for the use of organics in the petrochemical industry, it is also clear that these renewable energy feedstocks are not going away. For example, bioplastics have grown at an 8–10% rate per year, largely due to government subsidies in mature economies, such as the United States and Europe, but remain at less than 5% of the total beverage container market.[7] Yet, it does not appear that organic-based feedstock can achieve economies of scale and material design efficiency required for it to be competitive from a market dominance standpoint without greater government intervention. It appears as if the battle lines are being drawn between two enormous special interest groups in the United States: the petrochemical versus the agriculture (mainly corn) industry. Unfortunately, this becomes more of a debate amongst politicians and lobbyists over one incremental innovation rather than a competition amongst innovators to provide a disruptive innovation solution that will improve both the economy and the environment. If our 21st-century beverage containers are to be grown from the soil, they must be so from a science, technology, and supply chain standpoint rather than a conventional interest approach of a different type of happy cup fallacy.

Can these new organics grown from the soil replace the massive scale and supply chain of fossil fuels through a use of disruptive innovation? According to renowned chemistry professor Daniel Nocera, the answer is clearly no. Nocera states that the current world energy requirement is 16 terawatts, and assuming that energy efficiency can improve proportionately to the 50% of the current population that is underserved (a huge assumption!), a 50% population growth would leave an additional 16 terawatt energy deficit to overcome in twenty-five years, leaving a massive problem to address.[8] In Nocera's calculation, it is not possible for organics to solve this energy gap while also feeding a growing world population, especially if food production also takes a shift toward organics. While it may appear to be the solution to our fossil fuel problems of finite supply and environmental concerns, the math does not support this conclusion. Organics are only 1–5% of the entire petrochemical market and a far cry from a scalable alternative for energy and plastics. Even if a Material Scientist 3.0 was able to develop a radical innovation to achieve biodegradation and recycling for all beverage bottles in the future, it would be questionable, at best, whether many of these feedstocks that are used as food could scale up, given the other variables.

In a 2005 report by the Departments of Energy and Agriculture, it was determined that in order for bioenergy to have a significant impact (30% replacement) on replacing petroleum, a feedstock of 1 billion tons a year is required, equal to 25% of the current U.S. croplands.[9] Using 25% of our croplands to meet future world energy needs is not a realistic or even ethical solution to the problem; in the future, cropland will become even more important than it is today for both the United States and the rest of the world. Therefore, a corn-based plant bottle can never be viewed as a long-term scalable solution, or even a short-term solution for agriculture economics. From a supply chain standpoint, even the use of nonfood renewables is nowhere near the level of efficiency that would be required to achieve this 30% replacement value, which is critical from an energy/material perspective.

There is also an ethical dilemma in using food for energy or beverage containers in a world that cannot feed the present world population, let alone future growth that could lead to 10 billion human inhabitants. Finding sufficient food, water, and energy will become the greatest challenges in the upcoming decades, with solutions for each required to not suboptimize the others. While food surpluses currently exist in many of the developed nations, at least partially in the United States, a crisis exists in the developing world, to the point that the United Nations Special Rapporteur on Food called the use of food for fuel a "crime against humanity."[10] But food shortages are a problem in the United States as well: one out of five families faces problems of hunger, a statistic almost impossible to conceive. In the complexity of multiple market models within the United States and abroad, of industry, public policy, and special interests, the percentage of corn used for ethanol in the United States increased from 5% in 2000 to 40% in 2013.[11] The U.S. ethanol industry receives subsidies of $7 billion annually, while at the same time, the U.S. Congressional Budget Office reported in 2009 that the increased use of ethanol for fuel between 2007 and 2008 accounted for 10–15% of the rise of food prices.[12] Yet with the bushel price of corn at $4.32, which is only slightly higher than the actual production costs (between $3.95–$4.15), this appears to be a more complex problem than big business special interests; it is also a need to protect the financial viability of the U.S. agriculture industry.[13,14] Therefore, once again, the issue is never quite as simple as it may appear: on one side lies the concern of using such a large percentage of a harvest for fuel bioplastics while raising food prices during a time of low fossil fuel prices, and the other side lies a need to stabilize a damaged and important domestic production sector, even beyond an existing government subsidy program. To add to the complexity of the problem, consider that the United States loses 25–40% of its total food supply to waste, and this becomes a vexing problem to address.[15] It is no wonder that the 2016 presidential election will commence in Iowa with a large focus on corn ethanol subsidies; in March, 2015, Iowa corn mogul Bruce Rastetter moderated an Agriculture Summit where he invited the leading Republican candidates, with virtually all of them pandering to what the Iowan crowd wanted to hear.

FIRST GENERATION: THE INORGANIC ORGANIC BOTTLE

Even beyond the moral dilemma that the United States faces with regard to using a food feedstock for water bottles are the technical, economic, and environmental challenges regarding the legitimacy of using corn as an alternative feedstock for packaging. The amount of energy required in the production of corn ethanol is 91% of its net output, meaning that the net energy benefit from this process is less than 10%.[16] Corn ethanol yields 30% less energy per gallon than gasoline and is of questionable environmental benefit in regards to CO_2 emissions, as well as gas and soot. Due to its abundance as a crop in the United States, corn ethanol is used in the production of plastic bottles, such as Cargill's NatureWorks operation. Polylactic acid (PLA) is the substitute for PET, made from fermented plant starch (primarily corn).[17] PLA is touted as being "carbon neutral" since it is from a biotic (renewable) feedstock, does not emit toxic fumes when incinerated, and emits only two thirds the greenhouse gases of conventional plastic in production; however, when the entire supply chain is factored into the equation, the results are less spectacular, such as the 10 calories of oil required for 1 calorie of corn.[18] NatureWorks, the largest producer of PLA bottles in the United States, notes that corn plastic offers the most opportunities for environmental disposal, be it composting, incineration, or recycling. PLA is produced by naturally occurring lactic acid bacteria, which ferments sugars, and now with advances in fermentation methods is economically viable for packaging uses. Yet it is often questioned whether NatureWorks is a profitable business model, given the state of bioplastics in the United

States; in January of 2005, Dow Chemical sold its share of the company back to Cargill, noting a different expectation of consumer interest in such products.

Critics have concerns regarding the viability of PLA to conventional PET bottles. While PLA bottles will biodegrade, it is only under artificial conditions: a bottle will break down into carbon dioxide and water after three months of treatment at a composting facility under a controlled setting of 140° F. and plenty of microbes.[19] It can also be recycled back into lactic acid, but that requires a different process from other plastic bottles, adding complexity to the recycling stream. Many of these complexities are not well understood by the consumers who purchase them with the perception of a natural version of sustainability. The city of San Francisco purchased these bottles for use in conference boxed lunches, believing it was compostable; in the end, they stopped this program, finding the bottle to be logistically problematic from a composting and recycling perspective. According to a University of Pittsburgh study, bio-based feedstocks are not only questionable in regard to whether these materials are favorable from an environmental impact, but also whether these bottles could actually be worse in the end.[20] Despite these concerns, sales of the NatureWorks plastic bottle have taken off, with companies with green marketing programs like Newman's Own and Walmart taking the plunge. In Europe, the use of bioplastics has proliferated even more than in the United States, with its image of being environmentally sustainable leading to its growth beyond the ability of recycling infrastructure to process and reuse the material.

To break PLA into its components of CO_2 and water, an industrial composting facility with controlled conditions of heat (140°F), microbes, and lighting are required; even some of these facilities have announced issues in its breakdown. This, of course, is quite different from what the consumer expects when he purchases a bottle made of corn, so he may just throw away the bottle, expecting it to decompose by itself in a conventional landfill. In this normal landfill setting, PLA may decompose quicker than conventional plastic, but not by much; analysts estimate perhaps 100 to 1,000 years, meaning it may be a problem for fewer of your descendants, but it remains a liability that will extend beyond your natural lifetime.[21] These bottles are effectively indistinguishable from PET plastic bottles (other than the marketing) and can unintentionally contaminate the recycling stream, making the low rate of PET recycling (25%) and reuse (12%) even worse. At present, there are only 113 industrial-grade composting facilities across the United States, and a consumer base that does not know the difference between PET and PLA and does not really care leads to contamination within the recycling streams.[22] Recycling operations are already manual and fairly inefficient in industrial terms, so adding complexity to its process will only further degrade its efficiency, and thus material yields. Also, it is unclear whether it is better for the environment for a packaging material to partially degrade or not to degrade at all, given the issues with partial degradation leading to leeching, which can cause problems with consumption by wildlife. If PLA is one of the future materials for plastic bottles, it is unclear how a nationwide recycling strategy will need to be reformed from focusing on one primary material (PET) to numerous others. There is no evidence to suggest that corn as a feedstock can replace petroleum at the level of scale required and without the subsidization of the corn industry. Certainly, as a first generation bio-bottle, PLA has failed to fulfill its goal of being a true organic bottle. Perhaps the feedstock is organic, but the production process leads to the end product being an inorganic bottle!

Some environmentalists and some in industry claim that the use of a plant-based feedstock helps to reduce the amount of oil used to create plastic water bottles (which is estimated to be 17 million barrels of oil a year).[23] Certainly, there is truth to this statement, but, given that a barrel of oil consists of various carbohydrate possibilities that can be used for various types of products (such as auto and airline fuels as well as petrochemicals), it's difficult to suggest that a reduction in its use for petrochemical will reduce oil consumption unless the oil stays in the ground. Remember John D. Rockefeller's approach to reducing waste coming from a barrel of oil through the production of multiple by-products. We must also understand the net difference in energy and water use between a petroleum- and plant-based bottle, which is not very clear from extant literature. This should include the oil required to transport and process plant-based feedstocks; there has yet to be a valid measurement of its carbon footprint across the entire supply chain system, which is necessary to determine the most economic and environmentally sound policy. Yet a lack of clarity regarding the sustainability of these materials has not stopped companies from moving forward with their initiatives. In 2009, Coca-Cola created its Dasani PlantBottle, which included up to 30% fibers from Brazilian sugarcane and molasses and today is a 100% renewable bottle. Then the company launched its new Odwalla PlantBottle in 2011, the first single-serving beverage bottle made from 100% plant-based material. The company says it intends to increase the amount of plant-based material used in its Dasani plastic bottle by utilizing nonfood, plant-based waste, such as wood chips and wheat stalks to produce recyclable PET bottles.

Despite its name, the plant bottle has essentially the same chemical composition as a petroleum-based PET bottle, noted Christina Piles, who supervises the recycling of hazardous waste in Redding, California.[24] Ironically, even though these plant bottles are considered organic, toxic plasticizing agents are often required in its manufacturing, and its agriculture process often includes petrochemical fertilizers to grow the plants. "The compostable plastic industry started making this material without input from the composting industry," says Will Bakx, co-owner of Petaluma, California's Sonoma Compost. "They never thought through the lifecycle."[25] Although they may not be as natural as Coca-Cola would like you to believe, there are some upsides to these two bioplastic bottles. Both of them are recyclable, although there is some consumer confusion on this matter; the new Odwalla HDPE (high-density polyethylene) plastic is made from Brazilian sugarcane that is sustainably grown, rainwater-fed, renewable, and non-GMO. What's more, using this material doesn't take anything away from crops grown for human or animal consumption. That said, if Odwalla's HDPE bottle is made from such great material, why didn't Coca-Cola use the same stuff for other beverage bottles? HDPE can't hold carbonation, so it couldn't be used for soda, but why not Dasani water? One cynical explanation is the PR benefits of launching two types of plant-based bottles. Yet more likely, this is a supply chain network issue or a lack of scalability. In either event, this diminishes the likelihood of using any one type of biomaterial to replace fossil fuels as a feedstock. Are consumers as impressed with a 30% bio-based water bottle as with a 100% based juice bottle? There isn't clear consumer insight evidence. However, there's the technical challenge that I've presented already regarding an organic PET bottle replacement: most bio-bottles can be up to 30% organic due to the one monomer (MEG) easily fermented into ethanol from a starch source while the larger monomer by weight (PTA, or purified terephthalic acid) is much more difficult to replicate from natural sources. Yet, a plastic bottle can be 100% organic but not any more recyclable. All of this makes some skeptical: according to Jack Macy, commercial zero waste coordinator for San Francisco's Department of the Environment, "There's about as much greenwashing and distortion of reality as anything I've seen."[26]

Coca-Cola is also looking into the possibility of using PEF (polyethylene furanoate), a bio-based alternative to PET. PEF converts sugars to dicarboxylic acid, which can be reacted with ethylene to make the polyester substance. However, the impact of PEF on the recycling stream is still unknown; Coca-Cola is a major user of recycled PET, so using another material may harm the recyclability of the original PET material given the increased chemical complexity. Coca-Cola is presently working on bottles from PEF with Avantium, a renewable chemicals firm based in Amsterdam. Certainly, Coca-Cola and others in the beverage industry deserve credit for tackling this difficult issue in their efforts to reduce the amount of plastic that is released into our landfills. However, in the spirit of Man as Material Scientist 2.0, it is seeking solutions that either it doesn't fully understand from both a design and supply chain perspective or that cannot scale, therefore, adding complexity to the recycling system. Alternatively, it could be pursuing this with a marketing objective in mind. Perhaps it would be better served by gaining a better understanding before making a bad situation worse through greater complexity in today's already inefficient recycling system.

Another option is polyhydroxyalkanoate (PHA), which is made from fermented sugars or grown bacteria. It degrades much more easily than PLA but costs 20–50% more than petroleum-based plastic. All of these options are promising, to an extent, but appear to be only partial solutions that will lead to an all-of-the-above approach to bio-bottles. They cannot reach the scale of petroleum-based plastic, do not naturally degrade, and would lead to a logistical nightmare in the recycling system. Incremental innovations in material science piled on top of a supply chain system that is unable to handle the current levels of complexity do not solve the problem.

NEXT UP: GOOD SCIENCE?

What is important to understand from this chapter is that today's form of an organic bottle is produced using essentially the same manufacturing process as the one that creates long, beautiful polymers that are foreign to nature. Organic bottles have as much difficulty being recycled or decomposing in nature as a fossil-fuel feedstock bottle. Yet when we as consumers hear the terms *plant bottle* or *bottle made from corn*, we may instinctively picture a more natural, environmentally friendly container. Whether for noble reasons or greenwashing campaigns, these bio-bottles appear to be making the happy cup fallacy worse while impairing an already weak recycling system. While environmental advocates may be all too eager to jump off the fossil fuel-based plastic to something better, they'd better first watch where they are jumping to ensure that it's a better place to be. There appears to be little doubt in the research that, at least for now, the bio-bottle is not that better place. Hitting a single, at best, is no solution when a home run is required!

There is a lot of confusion in the public regarding what biodegradation means: Does it refer to ultimate biodegradability, hydro-biodegradation, oxo-biodegradation, mineralization (to CO_2 under aerobic conditions or CH_4 anaerobically), or compostability? There is also the question of whether the material needs to break down under optimal or normal conditions; under optimal conditions, plastics will degrade through biodegradation (bacteria), thermal (oxygen/heat), and photodegradation (ultraviolet, or sun and light). However, under normal conditions, the current processes to degrade plastic can range greatly. Then there is the question of how much break down needs to occur and whether it is safer to not break down at all, due to problems of leaching. Conventional plastics will partially and slowly degrade if left out in sunlight but not when buried in a landfill. Degradation is not like in nature; rather, these materials will break down to very small pieces, becoming dangerous entrants to our soils and especially our water systems where it is mistaken for food. As is illustrated in Table 5.1, some bioplastics are biodegradable, some are bio-based, and some are both; note that one of the bioplastics found to be both, polylactic acid (PLA), is not easily or truly biodegradable, as mentioned earlier.

Therefore, the problem isn't solely whether these plastics can degrade or not, but rather how long it takes, how much it costs, whether it occurs safely, and under what conditions it could be economically achieved within a supply chain system, including its impact on other more important sectors, such as food. Even if today's worldwide waste management

Table 5.1 Categories of Bioplastics

Biodegradable, not Bio-Based	Bio-Based, not Biodegradable	Both
PBS	PE	PHB
PCL	NY 11	PLA
PES	AcC	Starch

Source: Tokiwa, Yutaka, Calabia, Buenaventurada P., Ugwu, Charles U., and Aiba, Seiichi. n.d. *Biodegradability of Plastics.* Molecular Diversity Preservation International (MDPI). http://www.pubmedcentral.nih.gov/articlerender.fcgi?artid=2769161.

techniques were at a 99% effective rate, which is unlikely to be the case, there is the potential for tens of millions of these beverage containers to leech into our environment—often our oceans—on a daily basis, threatening it. Therefore, we must better understand the nature of these materials, organic or inorganic based, and its impact on our future. One well-known material scientist once told me the solution to the problem "is to place all plastic bottles from around the world in a large pit until we can best understand how to manage this material we do not understand today." Truth be told, nobody today has the full solution to the problem, regardless of whether the feedstock is from oil, natural gas, or corn. It seems as if Material Scientist 2.0 didn't fully think through the product lifecycle, either from an economic or environmental standpoint.

My fear with this organics craze is that there will be a rush to replace fossil-fuel plastic with bio-based plastics, an incremental innovation, without understanding the problem, further perpetuating bad science. Instead, we must better understand the real problems and solutions for every step of the product lifecycle, from design to reuse and disposal. Regardless of whether the bottle is made from petroleum, corn or bacteria, the question is not just whether it will degrade but rather how and how long. Will it be available for use again or at least not harmful to the environment? A world where we can be assured that all beverage containers are 100% reusable, biodegradable, or compostable is not on the near horizon as long as we keep plasticizing fossil fuels or organic materials like corn. Growth rates in use of plastic have been steady for the last decade while aluminum has been flat despite its infinitely better recyclability and reuse. It is easy to understand why recycling rates are falling due to the inability to recycle and reuse this PET material. In contrast, in order for recycling and reuse rates to increase, plastic as it is defined today needs to be replaced or transformed, and this includes any of today's organic-based plastics, such as PLA. Given the size of the petroleum-based plastic industry, and maybe even the corn industry in the future, this will be a huge policy challenge and will truly take revolutionary breakthroughs in material science in order to be achieved. The first generation corn or plant bottle does not meet that criteria. Material Scientist 3.0, where are you?

According to David Morris from the Institute of Local Self Reliance, the United States is the "Saudi Arabia of Carbohydrates," including nonedible lignocelluloses.[27] Lignocellulose is the most abundant organic material on Earth, accounting for 50% of all biomass. Suffice to say, this large mass of organic material presents an intriguing opportunity in the replacement of petroleum-based packaging, but quite a challenge in making this feedstock truly reusable or degradable. The major components of a lignocellulose material are cellulose, hemicellulose, and lignins. Cellulose, the most abundant biopolymer on Earth, makes up the plant walls of natural polymers of glucose and other sugars. Prior to the use of fossil fuels, it was these cellulose materials that were used as the natural polymers of manmade materials. Hemicellulose is the second most abundant biopolymer and has sugars easier to extract from plants than cellulose. These sugars extracted from plants are used as a basis for energy and material development. The last material is lignin, which constitutes the stubborn outside wall of the plant. Whereas hemicellulose and cellulose are more hydrophilic (susceptible to dissolving in water), lignin is more hydrophobic (water repellant), providing a protectant layer for the plant. Lignin strengthens the cellulose and hemicellulose, making hard wood, preventing animals from eating the celluloses (ruminants could), and allowing the plant to grow larger. Lignins become a problem when people seek to extract the sugars for use for bio-feedstock, such as the development of a bio-bottle. Yet what if plant breeding could lead to a molecular manipulation that would allow us to grow plastics or, perhaps more appropriately, grow plants that could be processed more easily into plastics that could be reused or would degrade easily in a supply chain process. We could grow our bottles from the soil and then return them back! The challenges in product design remain, as noted above, but solving this would become more of a game-changing, disruptive innovation for consumer packaging, rather than an incremental innovation with little benefit.

PHAs may be a better option for the future of bioplastic if it can be transformed in a manner different from conventional plastics. Typically, PHAs are conventionally created by supplying highly purified carbon sources, like glucose, leading to a very high cost of $14/kilogram, which is considerably higher than petroleum-based plastics.[28] However, researchers are looking into the creation of PHA from mixed cultures of industrial waste streams, such as soy wastes from a milk dairy or malt wastes from a brewery. Plant oils such as olive, corn, or palm are possible as a potential biofeedstock. Using cheaper oils as a renewable carbon source could improve the economic viability of PHA production. The greatest potential as it relates to the use of PHAs is that, unlike conventional oil and corn-based plastic bottles, it would be completely biodegradable under normal conditions. Therefore, it comes down to whether we can effectively transform nature in order to grow the right plastic. Only then can conversion processes that turn these biofeedstocks into a material that has the same properties as today's plastic bottles become possible. While it seems that materials like this could be the future of our beverage containers, the limitation we face today is our imagination to design and manufacture them in an efficient manner. At least with Material Scientist 2.0 at the helm!

Through the harvesting of natural polymers, such as rubbers, starches, celluloses, and proteins; the use of industrial biotechnology; and the use of transgenic technology (changing the structure of plants), we should one day have a water bottle that can be biodegraded and/or reused without toxic agents in the production process. Today, there appears to be a massive source of lignocelluloses from farms, fields, and factories that are waste materials, the type of stuff that John D. Rockefeller would find a way to reuse. Wood, straw, and agricultural and industrial residuals are not in as high a demand in the secondary markets as feedstock is in primary markets, as they do not compete in the food markets. Much of this waste is incinerated and landfilled rather than reused. Depending upon its classification, the size of this feedstock can be between 180 million to 1.4 billion tons,[29] a sizable volume to be used for bottle production and other material uses. Unlike corn that is grown to be used for PLA, secondary lignocellulosic material from farms, fields, and factories is not in conflict with more critical production/consumption needs. While some of these materials are used for feedstock, the sheer size of the material suggests that there is a sufficient supply for this and material reuse, like a plastic bottle.

In chapter two, I criticized the material science industry for the creation of the Frankenstein bottle, the PET bottle, and other creations that have led to a significant increase in waste and landfilling that has hurt our environment and economy. With an increase in beverage sales over the past decades being driven by these creations, packaging waste grows today, and our oceans are full of photo-degraded plastic pieces that marine life mistakes for food. Today, our consumer product companies are moving forward in a different direction, from petroleum-based containers to an incrementally improved plant-based one, yet there is little economic and environmental justification to do so at present other than for green marketing purposes. In many cases, the creation of yet another type of plastic that is difficult to recycle and reuse only complicates the problem, as we are just beginning to see in these facilities. So what does the future hold for us today in regard to this Plant Revolution, Green Economy, or whatever you would like to call it? Pervasive in this recycling myth is a theme of good versus bad science, business practices, government, and environmental advocacy intervention. On the horizon are some interesting prospects for biomaterials, but it appears as if Material Scientist 2.0 is unable to crack the code to balance economics and the environment. Perhaps we should take a step back from our enthusiasm and not make the same mistake twice. If we gravitate toward the convenience and greenmailing of buzzwords, we will get more of what we deserve, but don't want. If anything, we should learn the lessons of packaging success stories, like that of the aluminum can, that created a sustainable market of use and reuse rather than one that chooses economics over the environment, or vice versa. In the end, we will need breakthroughs in design over the entire product lifecycle in order to achieve our task. There are no shortcuts, regardless of whether the materials were grown from the soil or drilled from the ground.

Chapter 6

It's in Our Blood

TRASH TALK

Between Europe and the United States, there has been a lot of "trash talking," both literally and figuratively, in regard to the environment and recycling. Typically, it has been the Americans who have been the butt of European jokes, but maybe now it's America's time to have a laugh over the introduction the BinCam, a device to assist in recycling that is making social media even trashier. The goal of this device is to take a snapshot of your daily rubbish and then post the picture on Facebook for your connections to see. According to the developers, its social influence can change attitudes and behavior. On the BinCam's homepage (http://di.ncl.ac.uk/bincam/), you can see examples of how it works: points are awarded based on performance between friends, and you can even make comments (a post on Facebook, "Helen ate noodles again"). According to the leaders of this project, the use of social media to engage the young generation in recycling is the best approach to modify behavior—and to warn Helen of the risk to her heart of eating too many carbohydrates. Edward Snowden, where are you?

In comparison to Americans, most Europeans are willing to go great lengths to save the environment through their own personal behavior, even to the point of being ridiculous. But not everyone in Europe is on board with this traditional culture of environmentalism. Meet German chemist Michael Braungart, an ex-Greenpeacer who once lived in a tree in protest, but is now convinced that the use of regulation to solve environmental issues is counterproductive. Instead, Braungart believes that humanity's natural state is not to avoid or reduce growth through mandates but rather to resolve problems through progress. He notes examples of how going green and sustainable can become stupid human tricks: recycled paper that uses less virgin wood but contaminates millions of gallons of water, twenty years of paper recycling in Europe and no change to the toxic inks being used, and so-called green inks (e.g., soy based) that have a nefarious impact on the environment. In his words, "the conventional interpretation of sustainability is boring. It's about reducing, minimizing, and saving. In other words, do everything the same way as before, just not as badly. Not as badly is not the same as good, however. Instead of developing less damaging goods, we should be developing useful things."[1] In this book, I call out the use of milquetoast government, corporate sustainability programs, and organic feedstock in plastic bottles as examples of "doing less bad rather than good." From a chemist's viewpoint, Braungart provides a contrarian approach to environmentalism and the economy, after he himself was a true disciple! Noted innovators Peter Diamandis and Steven Kotler also find that future problems must be addressed through consumption and abundance in their book *Abundance*, subscribing to the role of disruptive innovation to address future challenges.

EU SUSTAINABILITY: SLOWING DOWN PLANETARY COLLAPSE

There is no question that the European Union (EU) is the most committed region in the world with regard to recycling programs. As shown in Table 6.1, the top five EU nations have recovery rates of over 90%, and there are a handful of nations professing "zero waste" (with landfilling rates under 5%). Sweden professes to be "almost there, with 99% of household waste recycled, one way or the other."[2] The difference between recovery and recycling rates, as shown in Table 6.1, is composting/digestion and incineration for "waste to energy programs." Of the top five EU nations of "zero waste" (Germany, Austria, Belgium, Netherlands, and Sweden), each of these nations incinerates over 30% of its packaging waste, and Sweden incinerates more than 50%! Under the EU Packaging Directive, all nations are required to achieve high reuse rates, defined as material recycling, composting and digestion, and incineration. Yet beyond the top five (Germany, Austria, Belgium, Netherlands, Sweden), and Lichtenstein, no other nation has a landfilling rate below 20%, and most nations have disposal rates consistent to or worse than that of the United States. With overall EU27 material recovery rates of 77.3% and recycling rates of 63.6%, the statistics are not as strong as

Table 6.1   Selected EU Recycling/Reuse Rates

Nation	Recovery Rate (%)	Recycling Rate (%)	Landfilling Rate (%)
Germany	97.4	71.8	<1%
Belgium	96.9	80.2	~1%
Netherlands	95.2	71.9	~1.5%
Austria	93.7	65.8	~3%
Luxembourg	93.0	68.2	~18%
Sweden	80.3	57.0	<1%
Italy	74.0	64.5	~26%
France	71.2	61.3	~28%
United Kingdom	67.1	60.8	~33%
EU27	77.3	63.6	~23%

Source: European Commission, Packaging Waste Directive.

often advertised; per the EU Commissioner of Environment, Maritime Affairs and Fisheries Karmenu Vella recently noted that, "waste is not managed as well as it could be," with a large gap between nations.[3] With the passionate "zero waste" nations incinerating approximately 30–50% of packaging waste, and over half the EU27 nations with landfilling rates higher than the United States, the results are less than truly sustainable. Furthermore, EU manufacturing firms spend an average of 40% of its costs on raw materials,[4] which suggests less of a circular economy than is advertised. Should a different approach be considered?

In this book, I have highlighted the difference between something that is *collected* as recycling and something that is *reused*. Most of us consider reuse to be either a bottle-to-bottle type of reuse or perhaps even an improvement in use, such as using an aluminum can for the making of an automobile or building. We do not consider reuse to include downcycling of a material, such as a glass bottle for road filler. As is shown above in Table 6.1, even the top five zero waste nations fall short in a like-for-like definition of material recycling, averaging anywhere from mid-20s to mid-40s in "downcycling" its waste to prevent landfilling. That appears to be fairly consistent with rates in the United States. These data support my hypothesis presented in chapter two: that recycling and reuse rates are capped economically and chemically based upon the material design and supply chain system of the container (i.e., higher rates for aluminum than PET), regardless of recycling practices across nations. For example, in Table 6.2, there is a similar pattern of plastic recycling rates in Europe as I presented from the United States, with collection rates twice as high as recycling rates. Recycling rates are rising, but no higher than 35%. While a 35% recycling rate for plastic packaging is significantly better than America's 10–15%, the use of costly mandatory recycling programs still leads to most of the used beverage containers being relegated to a "waste to energy" program, at best, and to landfills at worst. No EU nation has a recovery and recycling rate that would suggest a "zero waste" culture, despite such lofty claims. It is only through a large use of incineration as waste recovery that these nations can be viewed as environmentally friendly, even though the burning of beverage containers means that new ones need to be created from virgin material, and fossil fuels used for its replacement. This is no closed loop system, as some may envision. In the end, plastic as a beverage container was never meant to be recycled and reused, regardless of how hard an environmental culture seeks to solve the problem!

Despite such challenges in material design, Braungart's native country of Germany should be held in high regard for its focus on the environment; as is shown in Table 6.1, Germany is a near zero waste nation, with the highest material recycling rate in Europe, and less than 1% of its household waste is sent to a landfill. One of the first nations to develop a comprehensive national recycling program in 1991, its program is largely driven through an extended producer responsibility program (EPR) that holds the producer of packaging waste responsible in the process. In 1991, an ordinance was put in place regarding packaging

Table 6.2   EU 27 Plastic Recovery and Recycling Rates

	Recovery Rates	Recycling Rates
2005	50.9%	24.7%
2006	52.1%	26.9%
2007	56.5%	28.0%
2008	57.6%	30.3%
2009	59.7%	32.2%
2010	62.4%	33.2%
2011	63.4%	34.3%
2012	63.2%	35.5%

Source: Eurostat.

waste where retailers and packaging companies were responsible for collecting and processing used packaging from consumers.[5] Included in this regulation is a 60% recycling target for aluminum and plastic, and a 75% goal for glass,[6] which again, is much lower than would be expected from a zero waste nation, especially for aluminum, which is 100% reusable (at an 88% yield). In response to the 1991 ordinance, German producers created a *green dot* program that has become the model for programs across Europe. The green dot program has led to an increase in collection and recycling rates in Germany, but the program has been criticized for its high costs, unfavorable approach to waste management contracts in market, a higher focus on consumer packaging waste than commercial/industrial materials, loopholes, mismanagement, and a market saturation of secondary materials that are not in demand. As an example, the German Federal Environmental Agency is threatening to impose a tax on disposable bottles, given Coca-Cola's plans to increase its use; at present, 45.7% of all drinks sold in Germany are in a refillable bottle, and over 50% of Germans believe that a bottle with a deposit will be reused, which often isn't the case.[7]

Falling somewhere in between the zealous Germans and Swedes and the less than engaged Americans are the Brits. The UK is ranked as one of the top recycling nations in Europe at (eighth place) but woefully behind the top countries, given a landfilling rate of almost 40%. Contrary to the stern extended producer responsibility (EPR) programs in so many EU countries that make beverage companies fully accountable for their waste, the UK uses a purchase of tradable recovery notes (PRN), which essentially requires companies to purchase credits until recycling targets have been met. Indirectly, companies are held accountable for recycling targets established by the government, but a financial instrument guides how these targets are achieved on the market. As EU recycling targets increase, the price of the PRN increases as well, leading to more money at the disposal of the recycling companies in order to achieve targets. Inherent problems with this system, according to the Europeans, include transaction costs from additional parties, such as brokers that reduces fees for collection; the cherry-picking of collection of easier-to-achieve industrial packaging waste rather than consumer waste; the inability to use funds to invest in anything more than tactical collection; and the ability to point fingers elsewhere if the program fails, given the multitude of parties involved. Yet the UK has a higher material recycling rate than the Netherlands, which is one of the zero waste nations; the UK is deemed less successful due to its higher landfilling rate, which appears to be the benchmark in EU standards. Also, there appears to be trouble in the United Kingdom in regards to its program; recycling rates are falling due to *green fatigue* caused by EU mandates and the confusing bins imposed on citizens from recycling councils without a real culture for it. As a result, it is unlikely that the UK will hit the 2020 target of recycling half of its household wastes, leading to the country being subject to significant EU fines. The stalling of progress in PET plastic bottle collection rates, despite 96% of local authorities providing curbside collection, is evidence that EU recycling targets will not be met by these means alone,[8] placing the nation in an uncomfortable situation.

The developed nations of Asia, such as Japan, have similarly stringent recycling standards as the EU. Japan has a very mature and robust recycling program due to its being an island nation with few natural resources, a dense population, and a homogenous culture toward civic conformance. The task of sorting required of the Japanese citizen would not only be overwhelming to the most environmentally conscious American but likely too onerous for all but the most conforming Europeans as well! In Japan, meticulous rules and regulations are required of the consumer, and if not followed, there are warnings and penalties. Producers must also follow suit; as a result of the Containers and Packaging Recycling Law, all but small businesses are required to recycle all plastic, glass, and paper containers. Retail businesses have to collect containers on premise and pay for what is recycled by weight, leading to the prohibition of containers brought to be recycled that weren't purchased at the store! To the pro-recycling crowd, Japan is regarded as a success—plastics are recycled at a 77% rate,[9] which is two times higher than in the United Kingdom, three times higher than in the United States, and much higher than the most environmental nations in Europe. Yet beyond cultural conformance and resource limitations, Japan is blessed to have a synergistic opportunity with its large developing neighbor of China, which needs significant resources to fuel its production and consumption growth. Due to its geographical and political opportunities and challenges, Japan has an economic equation that enables a different approach to recycling and reuse.

China, of course, is a critical player relative to any commercial activity, given its global role as a manufacturing powerhouse with a growing consumer culture. For decades, this world's most populous nation of 1.3 billion has craved virtually any resource that it can get its hands on, including recycled beverage containers. According to Chinese custom officials, it received 7.8 million tons of scrap plastic in 2013—an 11% decrease—thanks to its Green Fence Program to reduce environmental damage in becoming the world's dump.[10] Used beverage containers from developed regions such as Japan, the United States, and the EU have been sent to China in mass, using this developing nation as a de facto landfill to complete its waste management strategy and achieve higher recycling rates. Yet China has its own recycling challenges from its growing consumer base. More than 20 million plastic bottles a day are thrown away in Beijing alone (within a underdeveloped recycling/waste management program), presenting another contradiction: How can a nation that imports recyclables also throw them away?[11] In Beijing, the Incom Recycle Company has built a factory that is able to complete bottle-to-bottle recycling, but its immature recycling system sells many of these materials to smaller manufacturing operatives, often in the textile industry, without proper environmental protection controls. In both domestic misuse and exported waste from the developed world, there is a growing sense of indignation over the environmental damage; Adam Minter, author of *Junkyard Planet* notes that, "China needs raw materials but nobody recycles anything for free."[12] Of every nation, China appears to be in the greatest need of a recycling program that matches consumption, given its impact on its economy and environment, and the environment of the entire world (particularly the oceans).

Differences in recycling collection and reuse rates are largely dependent upon resource availability (both in materials and landfill space). Any economic equilibrium model is a function of a nation's material resources; the composition and value of the packaging material (glass, aluminum, and plastic); the socioeconomic status of the general population; the maturity of its recycling and waste management systems; and the regulatory model from the government. Of these variables, only one factor is independent of national differences (material composition). Even if a nation, such as Germany or Sweden, has an environmental culture to recycle and conform to mandated rules, it is restricted by what can be done given the nature of the packaging materials and our global consumer supply chain. The market value of glass, aluminum, and plastic can only be incrementally impacted by recycling programs; chemistry, supply chain systems, and economics play a much larger role, as is shown in the net statistics.

ENVIRONMENTAL UTOPIA?

When an American recycling enthusiast lays his head down to dream, one place he may find his way to is Sweden. In a nation of 9.5 million people over 173 thousand square miles (versus 318 million Americans over 3.71 million square miles), the Swedes top the list of the most environmental people on earth. Unlike Germany, which has a smaller geography and almost nine times as many people, Sweden is a sparsely populated landmass that does not face the same waste management pressures; therefore, it recycles more out of culture than necessity. The Swedes are very environmentally conscious, cutting carbon emissions to exceed the Kyoto standards while at the same time achieving a similar economic growth rate as the United States, dispelling the myth that it has to be one or the other. Most Americans are largely unfamiliar with this nation, but it has become an important place for me since I began studies for a Ph.D. in a small Swedish town named Lulea in 2008, and later became a part-time post-doctoral researcher in Stockholm, the largest city in the nation. In general, the stereotypes that Americans carry about Sweden is that it's a place full of tall Viking men and women with high taxes thanks to socialism, low growth, and an environmental focus. Of these perceptions, the one about the economy is incorrect, as Sweden's growth in the past decade has been the best in Europe with a GDP growth of 12.6%, a rise in disposable income of almost 20%, and a public budget that is almost in a surplus.[13] It is this balance between the economy and environment that drew me to it for my research studies.

When I arrived in Sweden in 2008, I had carried same stereotypes in my mind. Arriving in Lulea, Sweden, a small town a little bit less than an hour south of the Arctic Circle, my initial impression was as expected: a beautiful landscape full of very content people, a laid-back work ethic (or perhaps better work-life balance than in the United States), and less of a consumer-based economic pull than I was used to. Yet the more time I spent in Sweden, studying the economic-environmental model, watching the average life of a Swede, listening to their perspectives, and seeing the results, the more I began to appreciate what I hadn't understood before. The Swedish principle of *lagom*, or *enoughness*, is an important aspect of not only their culture, but also how they consume and recycle. The model of *lagom* translates into not consuming too much and being responsible for what one consumes—believing that prosperity should be shared, not hoarded. Who can argue with these concepts? Yet in the back of my mind, I felt that something was wrong in this Swedish utopia. I just could not put my finger on what it was that led me to this suspicion. Maybe it wasn't something wrong with the Swedes but something that just didn't sit right with me.

Were my suspicions due to my American upbringing in a culture of a throwaway society? Or was it due to a veneer of a Swedish utopian society that was really much different on the inside? Neither of these explanations seemed to be the case. After all, I started to behave like an average Swede, consuming less, recycling more, and believing in *lagom*. There was a sense of pride when I recycled and processed my trash in the hopes of a zero waste model, mirroring nature. At the onset, I saw virtually all Swedes perfectly content to adhere to their culture; there did not appear to be any deference to individual desires of waste. After a while, the concept of the Swedish environmentalist became more than a curiosity, and it began to distract me from my core coursework. Like Mary Shelley, I was haunted by these thoughts, but didn't understand what it all meant or why I was so bothered by it. I could not quite believe a cultural approach that seemed too good to be true, yet everywhere I looked, it was there.

The feeling of intrigue was mutual. Many of my Swedish counterparts were curious about America as well, but for a very different reason: they were intrigued by our consumer economy but completely turned off by our excessive waste. After my studies had ended and I became a post-doctoral researcher, I presented an alternative approach to recycling at an Aluminum Recycling Conference Workshop in Trondheim, Norway, where most of the participants were from environmentally friendly nations like Sweden and Norway. In this conference, I respectfully acknowledged the Swedish success story but professed some concern with the data that I gathered and my experiences in an environment that created mandates and social conformity in the name of recycling. Just a few minutes into my presentation, I was confronted with my first contentious, uncomfortable showdown with a Scandinavian after spending five years in the region; abruptly, an elderly professor told me that my methods were mere gimmicks that were required only due to America's obsession with consumption and waste. I was a bit caught off guard at first, but I eventually gathered myself to question a model of sustainability that uses high conformance and mandates to mitigate a problem rather than to solve it.

From this confrontation, I realized what had bothered me about this Swedish environmentalism for the past five years. To quote the famous French philosopher Voltaire or, more recently, American author Jim Collins, "Good is the enemy of great." The effectiveness of *lagom* in the Swedish environmental-economic system could limit disruptive innovations that can really transform the discussion! To Swedes, this is a fair price to pay for the balance they achieve, but to use the words of Dr. Braungart, "Sustainability is only a tool to slow down the collapse." I came to the conclusion that had pestered me for so long that it eventually led to the writing of this book; for all of the effort being undertaken by the Swedes, Japanese, Germans, and others, it was simply an effort to make things less bad rather than actually fix the problem. After all, a material not designed for the environment, or even the overall supply chain, will never be anything more than a hindrance, regardless of how it is handled after use.

The history behind Swedish environmentalism is deep and strong; as early as 1938, Swedish author Ludvig Norstrom released reports titled "Dirty Sweden" that discussed the deplorable conditions in the Swedish countryside.[14] In the 1960s, the Swedish Society for Nature Conservation presented one of the first campaigns in the world titled "Keep Nature Clean," and Sweden was the first nation to establish an environmental protection agency in 1969.[15] Since 1947, the powerful municipal lobby Avfall Sverige has been a conduit for environmental sustainability in the field of waste management. Marketing itself as having 400 members and representing 99% of the Swedish population, Avfall Sverige has no American counterpart—an environmental lobbyist group as powerful, or more so, than private interests in the United States is a bit of a unicorn.[16] Yet in the past 15 years, Avfall Sverige has completed 360 development projects with a goal of educating the public and influencing public officials to achieve zero waste by 2020.[17] As a result, landfilling is being brought to almost zero, as it is defined by European (not nature's) standards.

In the United States, this type of regulation and special interest might be viewed as anti-growth, while in Sweden, it is viewed as a vehicle for growth. The Swedish parliament is enacting laws to cut CO_2 emissions, even below 1990 levels, as a measure that is both good for nature and the economy. Starting in the 1990s, the recycling and saving of materials was viewed as a method of improving the economy. Swedes are educated at the earliest years in both school and in home recycling programs that they are a significant part of the nation's largest environmental movement. Recycling is the action that requires the average Swede to partake in this culture of environmentalism; similar to Japan, consumers are required to clean and process recycling items in intricate detail prior to taking to a center. The level of commitment was more than just the act of cleaning bottles and cans; in 1975, the Swedish government passed the Recovery and Management of Waste bill, which mandated that the government pay 50% of the cost of recycling domestic waste.[18] These laws held the consumer responsible for processing, sorting, and transporting materials to recycling centers for processing, and held every citizen responsible for higher taxes to subsidize these operations. National and municipal laws were enacted, leading to commonalities and differences: in Stockholm, packaging must be taken to collection stations furnished by the FTI (Förpacknings- och Tidningsinsamlingen, or "Packaging and Newspaper Collection"), while Lund offers curbside pickup as well as the collection stations. There are also arrangements within large apartment complexes for packaging to be collected for the building for a fee.

Does this interlocking system of national and municipal laws lead to greater or less waste? If waste is defined as landfilling, it is obvious that Sweden is leading the world in the reduction of waste. There is also a collaborative culture between government, industry, and consumers rather than a contentious culture, as exists in the United States. Moreover, FTI is owned by five material companies in Sweden (Plastkretsen, MetallKretsen, Returkartong, Pressretur, and Svensk GlasÅtervinning), implying that the loop has been closed with regard to the integration of secondary materials into the primary stream, providing opportunities for producers to develop reasonably priced containers (albeit subsidized by consumers and citizens) in reusing secondary materials. Sweden's model is in line with its culture of *lagom*; its waste management model is an incremental innovation system, using subsidies through taxation and deposits to induce behaviors that reduce waste. Yet from a longer-term, strategic perspective, there is little progress toward the replacement of poorly designed materials that would enable transformational change; in fact, there is actually an overall decline in material recycling over the past four years, as is shown in Table 6.3! These results, after decades of focusing on environmentalism, actually show a stagnant end result and perhaps even an unintended discouragement of the innovation that is required to fix the waste problem. The Swedish model is a very effective Band-Aid, but this approach will never lead to economic and environmental growth and improvement.

The FTI has been accused of acting like Big Brother in its quest for perfect recycling program behavior from the citizens of Sweden. In Lulea, I had occasionally walked past some very dirty, unkempt recycling stations that led me to ask, what happens if you arrive to a recycling station, and there is no room for your materials? According to the rules that I was told, the citizen is responsible for driving the materials back home

Table 6.3   Waste Management Trends 2008–2012, Sweden

	2008	2009	2010	2011	2012
Material Recycling	33.4%	36.0%	33.7%	32.8%	32.3%
Biological Treatment	13.1%	13.9%	14.8%	15.0%	15.3%
Waste to Energy	50.4%	48.7%	50.5%	51.4%	51.6%
Landfill	3.1%	1.4%	1.0%	0.9%	0.7%
Total	100%	100%	100%	100%	100%

Source: Avfall Sverige.

and waiting for another day, which doesn't seem very environmentally sustainable to me (or economically sustainable, given petrol prices per liter in Sweden!). I was told the story of *sopspioners,* or trash spies, who watched your actions and could even get you in trouble with the law! People face Orwell's *1984* when leaving their materials next to a full bin, which can lead to it blowing all over the place, causing a mess, and being watched undercover as well!

Is Sweden an environmental utopia? It depends upon one's point of view: in comparison to wasteful systems, such as in the United States, it is very sustainable, but in comparison to a natural ecosystem, it is far from it. The advertisement of Sweden as a virtual waste-free society is a bit misleading, especially since the largest portion of its waste is incinerated, as is shown in Tables 6.1 and 6.3, which cannot legitimately be viewed as truly environmental. It also possesses a complex, if not bureaucratic, system of authorities that can be in conflict with each other, for example, FTI for material wastes and ReturPack for refund deposit schemes, plus the municipal waste system. The municipality waste company, such as Sorab in Stockholm, believes that it would be more efficient in collecting and processing packaging waste but is not allowed to be so, given the producer responsibility program. Perhaps a quarter of Sorab's collected waste could be recycled; therefore, it is potentially sending materials to incineration that are the same material composition of items that are recycled, after being washed out and stored by consumers. These organizations are bound within silos of legislation, which prevents competition and, therefore, can stymie innovation. Therefore, those who promote sustainability will find Sweden's system, and others like it, to be the best that exists today but woefully short of what must exist in the future. Yet I believe the Swedes are aware of the shortcomings of their own system, despite efforts from other nations to mirror it and proclaim it the utopia for the rest to follow.

WHY TODAY'S RECYCLING CAN BE GOOD BUT NEVER GREAT

To state my position clearly, the recycling models of nations like Sweden and Germany are good in comparison to the model in the United States, but insufficient to solve the problems as noted in chapter two. Therefore, any form of contentment in existing programs, or advocacy of new mandates, should be tempered relative to expectations. As you'll see in the solutions presented in chapter eight, programs need a little bit of influence from Stockholm, a lot from Silicon Valley, and even some from unknown innovators from crowdsourcing and prize-based competitions. Despite the high level of conformity in the Scandinavian nations in regard to recycling deposits being "in their blood," I am starting to see a generational rift in the younger citizens, as is also shown, less dramatically, in the United States. Both in Sweden and Norway, those same professors who challenged my radical innovation model to environmentalism quietly told me that the younger generations are not holding these civic obligations as tightly as their parents were taught to do, getting green fatigue. Gas-guzzling cars are making a comeback in the smaller towns, and even the Swedish Prime Minister Fredrik Reinfeldt was making a push for an EU-sponsored carbon emission cut to be tempered in order for those outside of his country to take on greater accountability, to carry their weight. Green fatigue is a faltering belief in environmental actions, a concern that the means are greater than the ends. Could the younger Swedes be having the same thoughts that "something is wrong," or are they becoming more individualistic? Households have been responsible for cleaning, sorting, and dropping off containers now for many decades. Could they be weary of "doing the same thing, just less badly," as Braungart diagnosed? Is there any hope amongst them for disruptive innovation amidst these interlocking bureaucracies?

Sweden's high priest of the packaging waste recycling refund deposit program, Pelle Hjalmarsson, the Chief Executive of ReturPack, noted that "it's in our blood to make deposits."[19] Something in one's blood becomes an unconscious act, which in the case of Sweden has led to short-term good but possibly long-term harm. In Sweden, a one Swedish crown (SEK), or kronor (worth 12 U.S. cents), deposit is placed on every plastic bottle and aluminum can purchased. This is refunded upon return; for plastic bottles over 2 liters, it's 2 SEK, and there may be administration fees on containers that are more difficult to process, such as colored PET bottles and plastics in general.[20] The centralized entity managed by Hjalmarsson is ReturPack, a pseudo-private company owned by the beverage companies and run at the direction of the Swedish Board of Agriculture and the Swedish Environmental Protection Agency. Through ReturPack, high recycling rates are promulgated through incremental innovations driven by government mandates that reduce the possibilities of disruptive innovations. This, in my opinion, could be viewed as a form of waste, at least as a potential opportunity cost.

In the United States, you'll be lucky to get the average American to participate in a comingled curbside recycling program, certainly not something of this level of complexity and effort. Matching the zeal in Japan, Sweden typically uses the following categories for separation:

- organic waste (mainly food)
- plastic
- cardboard
- paper
- metal
- glass (sometimes colored and uncolored as separate)
- batteries
- electrical
- hazardous waste (paints, solvents, medical)

It is not unusual for a Swede to have 10–15 different fractions of recycling. With so many recycling bins to manage, "garbage" is essentially nonexistent, but this can also lead to confusion. Also, storing all of these different materials under a sink is often not feasible, leading the consumer to store in other places, such as the basement, or to make more trips to the recycling center, which may be expensive, given fuel costs. In some cases, I have heard Swedes complain that their frequent trips to a recycling center offset any environmental gain in managing the material, perhaps leading to it being better just to throw these materials away. Generally, there are three waste processes in Sweden: recycling, composting, and incineration, with exported cans classified as recycled. The first, recycling, is the most known and regulated, with a great deal of expense and effort being undertaken to reuse materials. Plastic, glass, and aluminum used for packaging material are managed through a highly regulated supply chain system run by ReturPack, which is essentially a consortium company of the beverage manufacturers. These manufacturers are required to participate financially, and costs are passed on to the consumer. Because this program is funded by the taxpayers, consumers, and beverage manufacturers, there is little incentive for improvement beyond just incremental change. To further complicate the matter, materials flow to and from Sweden as a function of the EU, much like what happens in the United States from one state to the other. While this may lead to supply chain synergies within the EU, it also can lead to a specific nation's targets being convoluted or even pencil whipped in the process.

Another problem with the sorting of waste is whether it is considered packaging or not, which effects how it is recycled. If it's classified as packaging, then it has to be sorted and handled differently from something that is not considered packaging, even though it may be the same material. The reasoning behind these differences in procedures relates to who pays for the material (in the producer responsibility program, or EPR), but this makes the job harder for the consumer, who just wants to sort by material type. I believe it's theoretically possible for the same material to be processed as either packaging with a deposit and managed by ReturPack, packaging without a deposit and managed by FTI, or even non-packaging and taken to a municipal waste facility to be incinerated!

As shown in Table 6.4, recycling rates in Sweden for packaging waste have all achieved their targets; glass and aluminum recycling rates are very high due to the use of returnable glass bottles and to the value of the material, respectively; if aluminum cans are separated, it has a greater than 90% recycling rate. Plastic collection rates are high, but its reuse rates are much lower—as low as 7%, as shown in Table 6.4. Low recycling rates based upon the limitations of its material science is why ReturPack charges other fees in this deposit beyond the standard 1 SEK.

Table 6.4   Material Recycling Rates: Sweden

Material	Recycling Rate (2011)	EU Target
Glass	92.08%	60%
Metal	68.1%	60%
Plastic	25.97%	22.5%

Source: Eurostat.

Materials that cannot be recycled are incinerated for energy, which accounts for 50% of the energy for residential heating systems in the country. In the 1970s, Swedes started moving away from landfilling and toward incineration technology. In the 1980s, incineration technology was shut down for a period of time until the environmental impact was reduced. In 2011, the treatment of household waste was broken down as follows: waste to energy incineration (51%), material recycling (33%), biological recycling (including the creation of biogas) (15%), and land-filling (1%).[21] This high level of management by incineration may lead the average Swede to question why she is processing and cleaning all of these materials if the material is probably going to be incinerated. Should hot water be used to rinse out a material if it is not to be recycled? On the other hand, if it isn't, won't it smell when being stored inside the house? While Sweden has made efforts to reduce the waste associated with incineration, one has to wonder whether more than half of recycled materials being incinerated should be seen as an environmental solution at all. Every item that is burned for energy requires the linear supply chain system (extraction, manufacture, consumption, waste) to create a new product for consumer use, largely from virgin materials. Malcolm Williams, a director of the UK Zero Waste Alliance, believes that Europe will miss its waste-reduction targets for 2020 due to the high incineration rates in alleged zero waste nations: "It's just a myth that recycling is a difficult thing to do," said Williams. "So why on earth is anybody planning anything that is going to burn or bury more than 10 percent of the waste we're producing?"[22] While Williams's heart is in the right place, his science on the reusability of these materials may not be!

According to a 2002 study, the average American consumes 2.5 times the amount of soda consumed by a Swede, and that number is probably even higher today. In a Swedish supermarket, or *Systembolaget*, which is the state-run liquor store, beverages are typically purchased as single serve, whereas in the United States, they purchased by the 6-, 12-, or 24-pack case. This is entirely cultural, in my opinion. In the United States, I come pretty close to the average in daily beverage consumption (3.0 containers a day), while in Sweden, I dramatically reduced my use. I never thought it would happen, but I became a part of the *consume less* culture. But of course, for a Swede responsible for recycling, purchasing behaviors are affected by the type of container used, and where and how it must be recycled. Still, one has to wonder whether consuming less can lead to disruptive innovation, as has not been proven in other market circumstances.

IS BAD THE ENEMY OF GOOD . . . AND GREAT?

Is it too much to suggest that America, the land of disruptive innovation and lackluster recycling practices, could end up with a more effective long-term environmental and economic approach than Sweden in the future? Many find this unlikely; in a January 2013 CNN poll, the "environment" was ranked dead last in important issues, receiving only 2% of the vote.[23] In a recent academic study, 68% of respondents express neither a private value nor the perception of an external norm toward recycling in the United States.[24] As was noted in chapter three in addressing the throwaway society, waste continues to mount with little response from the federal government to control this problem; to find any significant efforts to reduce waste, one would have to go back to 1965 (the Solid Waste Disposal Act) and 1976 (the Resource Conservation and Recovery Act).[25] In regard to recycling laws, only eleven states have passed this legislation, and there has been only one law passed in the 21st century (Hawaii in 2005). Delaware abolished mandatory recycling in 2009. All of the other bottle bill states passed legislation between 1972 (Oregon, the first U.S. program), and 1987 (California), hardly the momentum the pro-recycling advocates are seeking.

In California, the California Beverage Container Recycling and Litter Reduction Act adds a nickel deposit on aluminum, glass, plastic, and bi-metal containers, and adds a dime for larger containers. On top of this, beverage manufacturers are charged processing fees for three of the four container types (plastic, glass, and bi-metal), with PET being charged $0.00012 and glass, $0.00192. The reclamation centers are reimbursed $0.0135 per container.[26] From one perspective, the program is a success because it has increased recycling rates from 52% in 1988 to an overall rate of 85% in 2013, including 100% aluminum, 85% glass, and 74% PET.[27] However, with a 20.7% administrative cost, the program requires a 79.3% collection rate in order to break even, which leads to a contentious balance between fiscal responsibility, taxation, and environmental targets. In 2012, an audit discovered that while 8.5 billion containers were sold in California, 8.3 billion were collected, which is a much higher number than CalRecycle reported;[28] the difference is the fraudulent refund of containers from other states, which significantly inflates the recycling numbers and creates funding deficits. Not only was the deposit being charged to the consumer and other fees charged to bottlers, the State of California was paying $10.5 million annually to cities and counties for the CalRecycle program, an expense that Governor

Jerry Brown planned to eliminate in the 2014–15 fiscal budget.[29] Amidst fraud, budget deficits, and collection rates beyond material market values, California's program is much like those in Europe: well-intended, but insufficient to solve the problem. Recently, California announced the "75% Initiative," with a goal to recycle 75% of all solid waste by 2020.

While bottle bill proponents are correct that curbside recycling programs collect only 23% of beverage bottles while bottle bill programs collect 80%, there is more to the story than, as Braungart notes, "making things less badly."[30] Take, for instance, another heralded U.S. bottle bill state, Michigan, and its highest deposit rate of $0.10 a container. While Michigan sets the standard in the United States with a higher than 90% return rate on all containers, based upon containers sold, more than $435 million in paper, metal, cans, glass, and plastic ends up in landfills every year.[31] The Hawaiian system also has concerns about a flawed payment system, inspection, and enforcement having an adverse impact on its financial viability.[32]

Is a Californian approach to recycling the answer in the United States? Given a lack of cultural support and questions as to whether this type of incremental innovation is effective in the long-term, it appears to be a Band-Aid at best, much like Sweden's longstanding system. So if not a Californian or Swedish approach to recycling, what then is the answer? Our attention turns to the great 20th-century economist, Joseph Schumpeter. While both Schumpeter and Karl Marx predicted that capitalism would ultimately collapse, Marx did so gleefully, considering socialism a savior, while Schumpeter found its demise to be a bad thing. Schumpeter contended that capitalism would collapse not because it is a bad thing, but rather due to being bastardized, with corporatism turning entrepreneurial activity into bureaucracy. Schumpeter dreamed of a capitalist system sparked by individuals with great ideas being used to continuously innovate. Creative destruction of the established would lead to new wealth. This was called evolutionary economics, a concept he created. In nature, the species that cannot adapt in the ecosystem goes extinct, and new species come forward and flourish. In this recycling story, some organizations and activities have promoted the same solutions for forty years without improved results, which in nature, would lead to their extinction. If Schumpeter were alive today, he would seek the emergence of novel forms of organizations and activities (speciation) and do away with those that perpetuate existing inactivity; this is the foundation of the solutions I propose in chapter eight.

This brings us back to the celebrity scientist, Dr. Michael Braungart. In his work, he creates products that not only mitigate environmental damage, but rather improve upon the environment. Imagine packaging that, after used, will naturally degrade under normal conditions and has seeds embedded, thereby helping the environment after use. For sure, the American consumer will not pay more for packaging containers as a function of environmental stewardship programs run by a bevy of interlocking government-run bureaucracies. While the environmentalists, and many of my European colleagues consider this to be a sign of selfishness, recycling has not made anything *better*; would the same American pay more for packaging that did? I think so. Americans need incentives, not mandates, to solve problems. Ironically, it is these mandated programs that lead to gimmickry, with innovations being the real solutions, despite some of my peers professing the opposite. And even worse, these incremental innovations can cause delay to or avoidance of the real solutions.

Disruptive innovation forces should address the problem of the difference between recycling and reuse rates. First, the differences between the container materials of aluminum, plastic, glass, and plastic-lined paper must be better understood, which is a focus of this book. My research finds that if only aluminum cans were recycled and not the other less recyclable packaging materials, reuse rates for recycled materials would rise dramatically. Perhaps the landfilling of lesser valuable materials seems to be a bad waste management strategy, but it will put greater pressure on the maker of these materials to design something better. There needs to be market equilibrium rather than the subsidization of poorly designed materials through over-recycling. From nation to nation, one thing is clear: a focus must be on waste reduction and innovation rather than recycling mandates and happy cup fallacy sustainability programs. Success in the future will not be measured by how much recycling is in a nation's blood, but rather innovation in its soul!

Chapter 7

Something Needs to Be Done!

If not a Swedish or California styled recycling program, what should be done to resolve this packaging waste problem? Even after consideration of the statistics presented in the last chapter, some will still believe that a European style nationwide recycling program is the answer, complete with a bottle bill policy and extended producer responsibility (EPR) program, while we wait for the disruptive innovations that many believe will never come from industry. Their reasoning is somewhat logical: Isn't doing *something* better than doing *nothing at all?* The statistics seem to support this statement, with nonregulated recycling rates in the United States hovering around 33% while mandated programs in the United States and EU are at 70% and higher. Why let 700 million beverage containers be thrown away on a daily basis, even if recycling them is a temporary solution?

Is a temporary nationwide recycling program in the United States while waiting for disruptive innovation the best option? Let's put aside the unlikely nature of this happening: in thirty-nine states, there is no recycling legislation and nothing even imminent. If anything, ground is being lost in existing bottle bill states, such as the 2014 referendum in Massachusetts that failed, with more than 70% of voters against it. While an argument in favor of a national program seems to be moot in 21st-century America, what if this changed, and there was political momentum toward such a program? In this chapter, I will predict what would happen if in 2015 an EU-style nationwide recycling program were successfully rolled out in the United States.

2020: A RECYCLING ODYSSEY

In this fictional account, there is just enough of a political base change in 2016 with regard to environmental and recycling legislation changes, leading to favorable federal legislation in the election year of 2016. A powerful national environmental and recycling political action committee modeled after Sweden's *Avfall Sverige* become influential in the United States, leading to the concept of a national bottle bill program and extended producer responsibility (EPR) program. As a result, a nationwide program is proposed for all bottles, cans, and cups purchased in the United States to carry a 10-cent deposit per container, with the beverage companies responsible for subsidizing the program. It is also proposed that this national program will be administered by a semi-private organization, regulated by the federal government to ensure consistency across state lines: no more fraud of entrepreneurs running cans from Nevada to California. Through the promise of environmentalism and job creation and the persuasion of an environmentally astute president, Congress passes a law requiring sweeping changes to the U.S. beverage supply chain, consumers, and the national waste management system.

Included with this bottle bill legislation is an EPR program that requires beverage manufacturers who bottle, import, and/or distribute in and to the United States to be accountable for its materials once they have been collected through this bottle bill program; materials are collected through a national refund-deposit system, administered by a federal department and/or semi-private organization, and the beverage manufacturers are then responsible for establishing a program through this semi-private organization in the hopes of reusing these materials rather than sending them to a landfill. The bottle bill legislation enacted in 2016 includes a two-year transition period for existing bottle bill states, and a four-year period for states with no laws in place. By 2022, the goal is to have a national recycling (collection) rate of 75% in comparison to the 2015 rate of 40%. In contrast to most current state bottle bill programs, this new program includes the increased challenge of all throwaway disposal beverage containers, including coffee cups, coffee pods, wine and liquor bottles, fast-food and convenience store plastic cups, and aluminum/plastic pouch drinks. As a result, this initiative will become one of the most ambitious recycling programs in the world to address the massive problem that America's throwaway society causes.

In this account, there may have been a (hypothetical) political platform for environmental and recycling program change in America, but the existing U.S. beverage supply chain system—including container and beverage manufacturers, distributors, retailers, consumers, recycling operations, and waste management companies—is not prepared for these challenges within a 2–4 year period, requiring too large of a logistical and cultural undertaking. As is noted in the last chapter, cultural conformance as a driving factor cannot be underestimated, and this often takes decades to build. Trying to retrofit an existing culture to a national program is much more difficult than starting a program from scratch with an existing culture in place, as occurred in Germany and Sweden. This is true not just from a consumer standpoint but an industrial one as well. Demand for beverages in the United States has a relatively high price elasticity, with pricing sensitivities being a significant variable to the overall supply chain system. While pro-recycling advocates contend that it is only fair for consumers to pay deposits to account for economic externalities, such as damage to the environment, this fluctuation impacts the market equilibrium (right or wrong), thus impacting sales.

As a result of these changes to the equilibrium in the supply chain structure, beverage manufacturers, distributors, and retailers must respond in kind to the impact from a nationwide bottle-bill added cost requirements across the supply chain in an industry of low margins. Right or wrong, changes that require these companies to be responsible for the collecting, processing, and reusing of these materials is a change to the structure of the existing supply chain system, and is not yet factored into their cost accounting systems. In Sweden, the price of a can of soda was two to three times higher, in my experience, than the same product in the United States when deposits and supply chain costs were factored in; it is conservatively estimated that a 10–25% price increase could occur in the United States if a nationwide program were implemented. This is supported in the data from other nations. As a result, market demand will fall, having a negative impact on beverage sales, reducing U.S. per capita consumption. Some may view this as equitable, or even good for society from a health and environmental standpoint. Many Americans will view this as an affront to individual liberty and the economy, as is evident from a lack of public support for such programs. With an unofficial U.S. inflation rate of approximately 6% (ShadowStats.com) and relatively flat labor rates, this discretionary consumer purchase may be impacted by a nationwide bottle bill program.

Not just the producers and consumers, but the states themselves will challenge the legality of federal legislation for this program. The Tenth Amendment of the U.S. Constitution explicitly states that the federal government is entitled to only those powers delegated to it by the Constitution, leaving a nationwide bottle bill program in question from both bottle bill and non-bottle bill states. The states argue that recycling programs are a state right under the Constitution, and they petition for the option to be exempt from a national mandate. Attorney generals from bottle bill and non-bottle bill states line up for a fight, promoted by consumer groups and the consumer beverage industry. Strange partnerships of non-bottle bill states and beverage companies form coalitions to bottle bill states wishing to keep these programs under state control and budget. Waste management companies also get involved, given their current business model that finds waste more profitable than recycling materials. America has less of a federalist culture of government than do many European nations, making it easier for special interests to challenge the constitutionality of such legislation. New entrants into a mandatory recycling program, such as coffeehouses, convenience stores, sport drinks, and other nontraditional beverages and retailers also raise questions, as they are completely unprepared for this change. They file injunctions to delay the legislation in order to study the impact of such laws on their supply chains and how it correlates to their existing sustainability programs. Wine and spirit manufacturers, who are outside the scope of today's bottle bill laws, petition for time to work through the ramifications of the law. Waste management companies file lawsuits in court, concerned that this sort of European-style law will eradicate their profitable business, as is the case in Sweden today. Higher recycling rates adversely impact their financials and waste management companies' ability to be the super-efficient mega supply chain that we rely upon to whisk away our trash.

In comparison to a culture of recycling nationalism that exists in Sweden and Germany, supported with a federalist approach from the start, an American nationwide program would be a retrofitted program likely attacked through lawsuits, injunctions, lobbying, and general dissatisfaction over such sweeping change on an issue that is not ranked high in priority by U.S. voters in most every poll. Take, in comparison, the health care debate in the United States, which is viewed by the public as a critical policy issue. Despite the importance of health care, the rollout of a national program has been fraught with complexities, lawsuits, and contentious national debate.[1] Is the U.S. public willing to undergo a similarly emotionally charged debate for an issue viewed as a much lower public priority? It is very unlikely.

After a two-year study of the merits of a national recycling program amid the pressure of lawsuits, injunctions, protests, and so on, Congress amends the law from being a federal mandated program to providing federal guidelines for the states to follow under their own programs. According to this revised federal law enacted in 2018, all states and private beverage companies are required to participate in a consumer- and producer-based recycling program with a target recycling rate of 75% in five years (2023). Instead of a 10-cent deposit on all drink containers, a compromise is achieved to require *at least* a 5-cent deposit, enabling states to charge higher deposits if they so choose. Another compromise is to no longer mandate distributor and retailer program implementation, but to require them to provide storage locations on premise for a third-party provider to manage these programs, paid for by the beverage manufacturers. States would be required to meet these standards by 2023 or would be subject to losing federal funding for projects such as highways and other infrastructure.

In this amended program, beverage manufacturers would remain responsible for funding the deficits of statewide programs in conjunction with each municipality, and for the collection and reuse of these materials; this becomes a logistical nightmare for beverage manufacturers and distributors, who would be responsible for up to 50 different regulatory arrangements mandated by state programs that they cannot influence. In an attempt to rein in the complexity and cost, the beverage industry seeks to replicate a privately funded consortium that combines the functions of BTI and ReturPack in Sweden to manage the process, hoping its charter to manage 50 different programs and state program changes is more bearable. Once again, challenged by the nature of state rights, there are some states that agree to this industry consortium while others choose to operate under a separate government agency, as is the case with today's bottle bill programs. Therefore, the implementation of a nationwide program with state loopholes becomes a logistical nightmare, leading to changes to the national supply chain network designs of many companies. In these models, the beverage manufacturers responsible for the containers after consumer use are also responsible for coordinating with its distributor and retailer partners while adhering to state laws and the national guidelines, being subject to audits from multiple parties.

By 2023, all states establish a recycling program, including a mandated refund-deposit scheme and extended producer responsibility (EPR) program because they risk losing federal funding if they do not comply. Between 2015 and 2023, when this program is finally implemented, there are been approximately 1.8 trillion beverage containers thrown into U.S. landfills, with no hope of reuse. In 2023, the first year of the national guidelines, there are variations between recycling programs, some of which relate to the state's commitment to participate in such a program and the role of the beverage consortium (versus state-run activities). With a few states choosing a 10-cent deposit instead of a 5-cent deposit, the 10-cent states programs appear to be on track for hitting a 75% recycling rate by 2025, while the 5-cent states are not on track. The 10-cent states run into funding deficits due to collection rates higher than 100% thanks to the flow of fraudulent materials from other states. In these states, the consumer supply chain is responsible for funding these discrepancies, further increasing the cost of beverages in the system. In contrast, the 5-cent deposit states are able to fund the program through lower recycling rates, unredeemed deposits, and container market value to cover administrative costs; however they do not achieve the targeted collection rates, especially with materials escaping to 10-cent states. Included in this mix are what the states pay for participation in this program and what is paid by the beverage consortium.

Eventually, the beverage manufacturers find themselves in a quandary of lower sales, higher prices, fifty different sets of regulations to administer, and the issue of many beverage containers being collected at higher rates than market demand. In Sweden, a national program is easier under very different circumstances: a smaller, more acculturated population with a lower per capita use; a national program from the onset; reuse of some materials (largely reusable glass and aluminum, and PET for lesser uses) in a highly subsidized model; and with higher costs accepted by consumers. The remaining materials are incinerated in a sophisticated waste-to-energy scheme that would not be market competitive with low natural gas and shale oil pricing in the United States. Plus, incineration always leads to more virgin materials used for replacement beverage containers. Also, there is no culture of returnable glass bottle programs in the United States, as is enabled in Europe by smaller geographical supply chains and government subsidization. Higher collection rates for aluminum in the United States are a win for the national program, and the economics of this program actually improves the price of secondary aluminum in comparison to primary aluminum, due to a more stable volume supply. One-use glass recycling and reuse rates also increase due to less commingling and thus lower damage and contamination, but this market has been eroding for some time now, given its higher supply chain costs. A national recycling program also stabilizes the supply of PET plastic, but market demand for secondary material does not grow, as is evident in low reuse rates in Europe. Private recycling entities, often run by the beverage manufacturers themselves, are put out of business due to these mandates. This weakens, not strengthens, the opportunity for reuse within the supply chains of these companies; the beverage consortiums are less efficient and are governed by state and federal laws. Finally, waste management companies lose revenue relative to lower landfilling volume, even though there isn't higher demand for secondary materials, which ultimately may be landfilled after recycling.

In this fictional account, a national recycling program, like exists in Sweden, Germany, and Japan, would bring some improvements, such as higher aluminum recycling rates and higher reuse yields from glass, but would be extremely difficult to retrofit to the United States, given a strong culture of states' rights, existing state programs, a lack of an environmental culture in the nation, the world's highest per capita beverage consumption, pressures from a super-efficient (yet wasteful) industrial supply chain system, and the massive supply chain transformation required to handle 1 billion containers used per day. As a result, a compromised version of a recycling program is enacted that proves to be too complex to be effective. Differences between states lead to higher collection rates in 10-cent states, but budget deficits and/or costs passed to beverage manufacturers. In contrast, 5-cent states have lower collection rates with budget surpluses, leading to a reward of the undesired behavior. Much finger-pointing and blame erupts due to the problems in this system, which impact recycling and reuse rates; also, it only adds to the divisiveness of pro- and anti-recycling positions. Not just in the differences between states is there a problem, but also in the supply chain systems of the beverage industry that wasn't in a position to transform, given lower sales and profit margins. In the end, it is a bigger headache than the nation is willing to tolerate, given its questionable commitment to recycling and the environment as a national priority.

SOMETHING NEEDS TO BE DONE!

Just because a mandated national recycling program is not feasible in the United States doesn't mean that the status quo is the best option. Can Americans, driven by a disruptive innovation model, define reasonable solutions that will enable them to consume what they wish, recycle, and reuse without wasting? That's what I will discuss for the remainder of the book. Would Americans be against this? I don't think so, as long as it makes sense for both the economy and environment; Americans are practical people, and they do not want to commit to a false choice of one over the other. Remember, while deposits may not be in our blood, innovation is in our soul!

My interest in this problem really started with my understanding of the aluminum can; as I mentioned earlier in this book, I have worked for the company where the aluminum can was first introduced for broad use in the beverage industry and in the development of the first recycling program that most of the beverage industry was against at the time. But I have spent the last few years trying to think through this paradox: How does an aluminum can possess such a low market value for an individual to recycle but is so much more valuable to a beverage can producer than virgin aluminum material? This is one of the ironies of the recycling myth presented in this book: a used aluminum can is actually worth much more than raw aluminum, yet the latter has higher market value to industry than the former! To financial markets (London Metal Exchange), primary (virgin) aluminum material fetches a higher price than this ready-made secondary material. A used beverage can typically trades about 20% less per pound than primary aluminum, a sign that the markets are irrational and inefficient. It just doesn't make sense to me that aluminum recycling rates are approximately 50% despite the high value of this material.

I followed both prices for months and found this price relationship between primary and secondary aluminum to not be an anomaly. I worked through the math with one of the world's leading metallurgists in a confidential setting, given his role in industry and mine in consumer products. We started with the price of prime aluminum on the London Light Metal Exchange, adding the processing costs in the United States (called "Midwest Premium"), and calculated a UBC price per pound. Then, we determined the number of cans per pound (roughly 34), worked through the processing costs through the supply chain (collector to dealer to consolidator to broker to smelter), and determined the recovery rate (88%). Given that the reuse of UBC requires virtually no additional alloying, this was the cost of processing UBC for reuse as an aluminum can. In contrast, we calculated the cost of processing virgin aluminum at market prices to that same aluminum can composition. To do so, there are five metals that must be added for alloying, and there is a loss of materials in the process as well. Depending upon whether the producer using virgin aluminum was a manufacturer itself (like Alcoa) or a large customer of a manufacturer, we came up with costs and compared them to the reuse of UBC, as is shown in Table 7.1.[2]

Table 7.1 is evidence that our commodity markets are out of sync in regard to the value of primary and secondary materials. How can it be true that a more valuable material is traded at a lower price? More than

Table 7.1   UBC Reuse vs. Primary Aluminum[3]

	Aluminum Manufacturer	Large Aluminum Buyer
Cost per LB difference to UBC (in cents)	+4.06	+11.96

Source: Buffington and Peterson, 2013.

likely, it's due to our massive supply chain production systems that are designed to link to virgin, raw materials rather than secondary reused materials in a closed loop model. In effect, this is the waste that John D. Rockefeller challenged his managers to reduce as a matter of economic and religious principles. Yet today, and every day, tens of millions of aluminum cans, if not more, are tossed into landfills to be buried as worthless trash when they have high market value. Something needs to be done about this!

Today, plastic bottles have been found, in some studies, to release small amounts of these synthetic materials into our bodies that may mimic estrogen for women;[4] the potential of this problem is largely debated, and it seems as if there is no conclusive evidence on either side, especially given the relatively short era of plastic and its long half-life in the environment and our bodies. We do know, however, the impact that plastic is already having on our ocean and other marine systems. When I was growing up, I remember the allure of the message in a bottle, of going to the ocean, putting a note in a bottle, and hoping that it would arrive in Europe for someone to open. Today, our oceans are full of bottles, bottle caps, and other assorted plastic that somehow escaped our waste management systems, given the magnitude of the materials to collect and process. While you may not see it in our oceans, microplastics, which are 0.33–0.99 millimeters in size find their way into waters and animals and account for 81% of the total trash in the Great Lakes.[5] There are five ocean garbage patch zones consisting of large masses of plastic, some larger than the state of Texas. Some estimates place the problem to be up to 3.5 million pieces of plastic per square kilometer of ocean![6]

These estimates are based upon today's impact on the environment; if its growth continues for hundreds, if not thousands, of years in the environment, the future could be devastating to the natural ecosystem. Today, microplastics exist in the marine systems for wildlife to consume, and submerged plastic is relatively undetectable. Marine biologists paint a picture of a seabed full of plastic that reduces oxygen levels, upsetting the balance of oxygen and carbon dioxide, effectively destroying the chemical nature of our most valuable ecosystem. In 2014, this problem of plastic trash in our oceans was highlighted during the search for the Malaysian Flight 370 jetliner when trash was mistaken for pieces of the plane. Another reflection of this is the 100,000 mammals and millions of fish killed every year. But it is not restricted to oceans; the Thames River in London has been considered clean after a checkered past of pollution, but a recent 2012 study found thousands of plastic relics scattered on the river bed.[7] Floating on top or unseen deep below, the sheer volume of plastic in our society seems to be almost unsolvable. And this does not even address the potential health impacts of plastic to our own bodies; a recent study found that 75% of plastic release estrogenic chemicals.[8]

Recycling programs can only reduce the bad and distract us from fixing the problem. This is the case in the developing world and the less sustainable developed nations, such as the United States, but also in the environmentally friendly ones, such as Sweden. The issue is not whether we should encourage or discourage consumption, or whether economics or the environment is more important, but rather, the reduction and elimination of waste, and finding an incentive to use more materials that are good and fewer that are bad. Waste can take on many forms: overuse is one, but the collection and reuse of materials in a forced fashion is wasteful as well. A policy to protect the environment at the expense of the economy will achieve neither while a policy to balance the two will accomplish both! If anything, our recycling programs have aimed too low and achieved even less. A target of zero waste must be compared to nature, not a wasteful industrial supply chain. Consumers and companies should be incentivized for doing the right behaviors rather than penalized through mandates regardless of their actions. We must reduce waste through innovation, not regulation.

In the next chapter, I will lay out solutions addressing this packaging waste problem by balancing our needs for economic growth and environmental sustainability.

Chapter 8

Solutions—Extending the Supply Chain to Nature

If you do not change direction, you may end up where you are heading.
—Lao Tzu

A PACKAGING REVOLUTION?

Eventually, an economy fostered from waste will run out of steam; when will the negatives of the throwaway society outweigh its benefits? In this book, I have made the case that we are reaching this threshold, even if we do not know that it's happening; our mega-efficient waste management system and conventional recycling programs are able to whisk away the problem to make it temporarily invisible. The problem troubles both the environment and economy, even if we are not noticing the impact. Yet, this waste is completely unnecessary. A new socioeconomic and techno-logical system could spur a packaging revolution driven by good science and sponsored by industry practices, leading to good jobs for the young and consumption that can change the culture rather than bringing guilt with it. A plastic bottle of the past and present is a marketing device, a lightweight and sanitary method of growing sales through portable con-sumption. What will the container of the future represent?

The five solutions that I propose in this chapter are entirely possible if strategy is redirected away from mandated legislation and lackluster corporate sustainability programs and toward transformative change. These five subsegments of an overall disruptive innovation model bring together stakeholders (consumers, producers, advocacy groups, and governments) into a collaborative model, breaking down ideological differences through a focus on the ends rather than the means. Rather than copycatting the best conventional recycling programs (incremental innovation), a new platform can be established for something much better through disruptive innovation real change. There is no place better to achieve this packaging revolution than in the United States, given its high consumption and waste, unwillingness to further government mandates, and entrepreneurial and innovation-based spirit. So far in this book, I have articulated what hasn't worked; in these last two chapters, I will outline what must be done for this new system to be put into place.

SOLUTION 1: MATERIAL SCIENTIST 3.0—A (SUPERCOMPUTER) CREATOR OF NEW MATERIALS

It's time for producers and consumers to stop the use of throwaway, one-time use beverage containers that were never designed to be recycled/reused or biodegraded in the first place. A correction of today's recycling myth must begin with not only how these containers were designed in the first place but also their supply chain systems. For example, not only does a supply chain of reuse exist for aluminum cans in the consumer beverage industry but also in the automotive, aerospace, and building sectors. In contrast, most of the other materials aren't even reusable within their own markets, having to be downcycled at best in order to avoid landfilling. This is a problem of design, and I believe that it is fair to say that today's mandatory bottle bill programs have been focused on the wrong paradigm for over forty years. Therefore, the material design paradigm must be turned inside out; instead of designing the container for part of the supply chain and then trying to determine reuse after the fact, the design of the material must optimize both front-end and back-end recycling/environmental needs, potentially with multiple systems of use. Setting this as a design objective requires that the consumer beverage supply chain, including recycling programs, look at itself in a much different manner. It also enables others to bring innovation from the outside to create solutions.

In chapter two, I defined Man as Material Scientist 2.0, the second designer able to create things, after that of Nature/God, who is Material Scientist 1.0. The main limitation facing Material Scientist 2.0 is a temporal problem: Material Scientist 1.0 has the luxury of millions of years of emergence and self-organization to create materials that are in synergy within an ecosystem. When humans created the PET plastic bottle in 1973, they invented a synthetic material over a couple of decades that is foreign to organisms. It was optimal to the front-end supply chain but not nature. Without symbiotic relationships formulated through ancient calculations, today's industrial materials are designed outside of a closed loop system, which inevitably leads to conflict with the environment, as we are seeing in our oceans and possibly even our own bodies. No matter the efforts on the back-end, plastic cannot be efficiently reused or degraded within the existing supply chain systems and, therefore, follows an unnatural process during landfilling or littering called leeching—a nasty afterlife that allows microplastic to enter the food chain insidiously for birds, marine life, and possibly even ourselves. Paradoxically, material scientists have been God-like in their ability to create products of use to transform our lives, but infantile relative to their knowledge of its cradle-to-grave implications. The reason for this is clear: our limitations as designers prevent us from optimizing beyond a few variables, which leads to limited, albeit convenient creations. Since we cannot design products over thousands of years, enabling self-organizing relationships to form, perhaps we can utilize the powerful computational abilities of supercomputers to calculate supply chain and ecological symbiosis in a few months or even weeks. While perhaps not achieving the beauty of nature in a perfect manner, the use of technology to bring together industrial and ecological materials and systems could lead to suitable solutions that tie industry to nature.

The limitations of humanity's ability to design in order to extend the industrial supply chain to nature is obvious through such inventions as the PET plastic bottle and today's inorganic "organic" product designs, as discussed in chapter five. We have nothing more than incremental innovations to slow down the pace of economic and environmental damage, unable to transform our supply chain systems to a balance with nature. In the future, supercomputers should be tasked with an exploration and calculation of thousands of permutations of material compound combinations that are not possible through human calculation. The solution should not only optimize specifications within the front-end of the supply chain—such as manufacturing, shipping, and consumer requirement—but also reuse logistics, processing, remanufacturing, biodegradation, and ultimately nature. Just as mathematics provides the raw computation of life on Earth, supercomputers can replicate a similar process,—not over thousands or millions of years, but rather days, weeks, or months; in comparison to today's Edisonian method of innovation, this would be a radical change in the field of packaging design, if undertaken. Material Scientist 3.0 will be tasked with solving a different problem: to create completely novel materials, perhaps from elements never considered for industrial packaging use or from compounds that never existed in the past or present. Through raw, exponentially superior processing abilities, researchers (not material designers) will begin to test thousands of materials in relation to an overall supply chain system that includes nature, and to eliminate infeasible options in the virtual world before wasting a minute or dollar on physical prototypes, testing, and marketing. Therefore, the supercomputer will not only create the product and the supply chain but also the product's relationship with nature in order to make nature glad, leading to a new paradigm for industrial growth. Green chemistry, powered by the supercomputers of Material Scientist 3.0 to the rescue!

For those who think that Solution 1 is nothing more than science fiction, consider the revolution already occurring in the field of material science, at least in basic research and public policy, such as the Material Genome project. The goal of this project is to change the product innovation timeline from an average of 10–20 years to 2–3 through three innovations: computational tools, experimental tools, and digital data. All elements, natural and synthetic, will be digitized according to its material specifications, just as is occurring with our human genomes. By encoding information to material elements and compounds, product design parameters can be number crunched using powerful supercomputers in order to create perfect combinations of physical design and supply chain processes; self-organizing relationships can be virtually modeled before physical prototypes are attempted. For example, a beverage manufacturer will be able to establish the specifications required for the container, and then a material scientist, through the use of Material Scientist 3.0, will be able to compare hundreds or even thousands of parameters to potential material compositions, whether they currently exist or not. A supercomputer will develop existing and novel compounds to consider for this functional use and will redesign supply chains accordingly. In an open source networking environment, researchers from all around the world will be able to collaborate in the search for material innovations. Using supercomputing power and crowdsourcing, hundreds or thousands of material combinations can be virtually tested in a rapid manner, skimming down a list of potential material options in a short period of time and at low cost. The opportunities for innovation will be practically endless and will be built upon digitized data to define these critical beverage containers of the future. Afterward, 3D printing and nanomanufacturing techniques can be utilized to expedite the product to market. A balance of time to market and material design/supply chain thoroughness will be achieved!

Think about the possibilities—the potential to create new material combinations never considered before and to achieve oneness with nature in the creation and disposal of these materials. Perhaps relationships will be discovered between new packaging materials and bacteria in order to enable the breakdown of the beverage container, or some other symbiotic relationship will be found. Perhaps the creation of programmable matter—where a beverage container exhibits industrial specifications upon use and then transforms itself into something harmless or symbiotic with nature—is feasible. Can conventional plastic materials be redesigned for safe and efficient use in the environment? Maybe not, but perhaps in using this process, the PET plastic bottle can be reverse engineered from both a material and supply chain process to achieve 50% reuse, then 95%, or even 100% biodegradability. A new design paradigm can analyze the feedstock (abiotic or biotic), chemical/manufacturing process, distribution, and post-use and ecological system to ensure the end result meets industry and environmental specifications. While it may sound preposterous to outsource so much of the material design away from human designers to supercomputers, we have to consider that nature is built on an endless array of mathematical calculations over thousands and millions of years, a process that humans can never come close to replicating. Our own bodies are made from DNA that many scientists believe is akin to computer code; while supercomputers may never achieve a level of sophistication to calculate the creation of a human, it seems entirely possible for it to create a higher order of beverage container and other consumables.

A material genome approach to developing transformative materials beyond the dreams of human inventors (supercomputers as Material Scientist 3.0), will enable these new materials to not only completely and radically change our supply chains of the future but to extend the supply chains into nature! Transformation will be unleashed on the extraction, design, and procurement of new materials and then to its manufacturing, before being reused or even deposited back into nature. Instead of a material designed for front-end use only, these new materials will be optimized from cradle to cradle, as Michael Braungart has raised in his challenge. It's difficult to conceive how our massively wasteful consumer beverage supply chains can be transformed in such a manner to not just enable economic growth but also to make the environment better in the end. After invoking this transformational approach to material design, we will look back upon our 20th century design, manufacturing, logistics, and recycling/reuse practices as painfully remedial, not having the understanding of how accelerating technology could solve these economic and environmental challenges. Perhaps packaging containers can be produced at retail through 3D printing rather than at a manufacturing location in order to reduce the material requirements, making it easier to decompose. There's so much potential and no time to waste! The greatest challenge will be for Man as Material Scientist 2.0 to let go and understand that our limited ability to design will never achieve a solution to the level that is required to optimize the economy and environment in a developing world of more than seven billion people. Every day that we wait to undertake this disruptive innovation will lead to over a billion wasted packaging containers in the world.

SOLUTION 2: GOOD OR NO PACKAGING THROUGH A NEW APPROACH TO INNOVATION

With Solution 1 in place, the future of packaging will be transformed by a new practice of green chemistry, material design, and supply chain systems to be extended to nature in a profitable, growing manner. New materials and designs, as well as redesigned supply chain systems to fit the purpose, will lead to efficient, recyclable, reusable, and/or biodegradable packaging products that grow the economy and stabilize the environment. This will also transform our definition of innovation: instead of solutions being solicited through a search for individual, creative entrepreneurs from out of the blue, or giving research grants to institutions, the radical innovation will have a mathematical foundation, and the innovators will arise from crowdsourcing and prize-inducement competitions to optimize industrial supply chains to ecological systems. Therefore, transformation won't arrive from a dream of an individual sitting underneath a tree and getting hit in the head with an apple, or even from isolated labs and institutions. Rather a 21st-century approach will combine superprocessing computing power and the power of crowds across the globe to create symbiosis as powerful as nature, or close to it. Today, we think of a 95% reusable or 100% biodegradable container as impossible, and we modify our definitions of a zero waste system to make it achievable. Yet imagine a new model of innovation that spurs collaboration, not just within a crowdsourcing scheme of like-minded material scientists across the world, but by linking supply chain partners (beverage and container manufacturers, distributors, retailers, and consumers) and even government bodies, environmental advocacy groups, and virtually anyone in the world with the best ideas. In nations such as Sweden, there is a whiff of collaboration through these mandatory programs, but it mandates agreement rather than inducing innovation. Exponential improvement in economics and sustainability cannot be constrained through legislation or boundaries of any sort; it must be triggered through competition-based innovation and crowdsourcing. Today, there are technologies in place to get us there if we set our sights beyond our conventional wisdom from the past forty years.

Now is the time for our model of innovation to change, for us to be practical and learn more of how creativity has happened in nature via complexity theory of nature-based algorithms, including biomimicry and nanotechnology. As we understand the intricacies of nature's designs and materials, such as in nanocellulose, we can get insight into how we should design the material and supply chains of beverage containers, not just in using organics, but organically inspired supply chain processes to truly replicate nature. A movement toward bio-materials that don't conflict with our food supply must also be complimented with the bio-supply chain of materials, including extraction, chemistry and manufacturing, distribution, and even redistribution back into nature. Wood pulp, corn stalks, spent grains, and even bacteria must be used efficiently and reused rather than being left outside of our mega-supply chain systems, as is the case today. Catalysts and agents to achieve chemical compounds must be safe and nontoxic. The key to this transformation is in nanomanufacturing, the concept of designing, manufacturing, and distributing in the range of 1–100 nanomaterials (with 1 nanometer equal to one billionth of a meter in size), supported by Solution 1. Imagine the use of nature's waste to create beverage containers that can be returned to nature as biodegradable waste for use in an ecosystem. This could lead to interesting, and not forced, symbiotic relationships between farms and packaging manufacturers, with the waste of the end product going back and forth between factories and fields! Only through a different approach to innovation will this be possible.

Supercomputers and crowdsourced researchers will also develop new packaging that can be *programmed*, so to speak, to become reusable after its initial use or fully biodegradable; in this manner, a beverage package will possess a certain form and function, only to be transformed into a new after-use material that can be readily used as a feedstock for a packaging material production line or discarded to nature. Future materials will have this ability to change form and function so that the same material can serve varied functions in the overall supply chain. In this scenario, two primary goals can be achieved: first, an existing packaging material can have a high yield of reuse or biodegradation from the original container (my goal of 95% and 100%, respectively), and second, the manufacturing process will require less energy and toxic materials in remanufacturing. As Janine Benyus notes in her famous book *Biomimicry*, industrial processes require enormous amounts of energy through the processes of *heat, beat, and treat*. These new programmable materials will dramatically increase the reuse/biodegradability rate of materials to 95–100% through supercomputer designed compositions, while at the same time using less energy during the transformation from used to new or natural waste. In the future, we should never consider the burning of a plastic bottle in a waste to energy scheme to be sustainable for the environment. A paradigm shift in product design and supply chain systems should be driven via a new approach to innovation.

What if packaging waste was optimal rather than dangerous when littered? What if your water bottle could not only be safely thrown into a forest or ocean, but was actually good for the environment when you did so? Through programmable matter and other transformations not yet known to us, our industrial materials post-use must be safely transformed from supply chain compliant to fish and bird food or other natural ecosystem materials, completing a cycle where consumption would lead to good, not just less bad. Our industrial activity could become a natural extension of the ecosystems of the earth! Never again would we have to see a dead sea bird on the beach, full of plastic.

Examples of other known innovations on the horizon in beverage packaging are the design of *aerogel*, a material that is approximately 2% solid material, with the balance being air, with particle sizes of 2–5 nanometers. Made from carbon or other materials with the appropriate properties, this material of less organic substance may lead to less waste in packaging while offering even better supply chain performance than we enjoy today. Another possible innovation could be the combination of nanomaterials and 3D printing in order to embed solar cells in packaging that would cool the beverage without the use of refrigeration! Techniques in intelligent packaging and printed electronics offer transformative opportunities for our packaging of the future, but the conventional consumer beverage supply chain system must look in different places for such innovations than it has done in the past.

A new approach to innovation must include entrepreneurial start-ups via crowdsourcing; prize-based competitions, such as the X-Prize; design via supercomputers; manufacturing and distribution via 3D printing; and postponement that translates into recycling and reuse. This will lead to the most economic and sustainable packaging of all: that of the *low or no packaging* product. This will change not only how packaging is designed, produced and distributed, but the beverage as well. Today, our beverages are mainly packaged where they are manufactured and then shipped, possibly thousands of miles, through the use of various fossil fuel-based transportation vehicles, such as tractor trailers, railcars, and ships. If supply chains are designed as is nature, beverages can be *postponed* in the supply chain process to the retail point of sale, dramatically changing the type and quantity of packaging required, including a reduction in secondary (e.g., corrugated paper) and tertiary (e.g., pallets and shrink wrap) packaging; perhaps this would lead to a different design of packaging material that is more likely to be biodegradable. Simple manufactured beverages, such as waters, soft drinks, coffees, and juices could be transformed through its supply chain as much as through packaging; fermented beverages that require longer manufacturing processes may require different challenges. A low-packaged product could be achieved very quickly and be enabled through the emergence of 3D printing techniques that will transform our consumer supply chains all by itself! Changes must occur, not only in the design of the packaging but also in the supply chain. Through the technology of crowdsourcing and prize inducement activities, radical change will arrive at the consumer beverage industry sooner than we think possible!

If we head down this path, there is no question that green chemistry, nanotechnology, 3D printing, and supply chain postponement could revolutionize the beverage packaging industry of the future. Technologies will be in place for this from one end of the supply chain system and extended to nature, but this must begin with a different model of innovation that spans beyond publicly funded research and corporate R&D departments to also mirror nature: self-organizing innovation as opposed to pre-determined solvers. The generation of solutions must extend beyond a few designated creative types with a baseline of the computational power of supercomputers, open sourced knowledge, and crowds that form complex and chaotic relationships through unique incentive programs. Every day that we expect corporations, academia, legislation, and government policy to solve the problem, or that we wait for an individual genius with a dream to solve the problem, we further degrade the economy and environment. Solutions 1 and 2 are the foundation of transformation that can drive economic growth through a packaging revolution.

SOLUTION 3: A VOLUNTARY REFUND/DEPOSIT SYSTEM

With the process of how materials and supply chains should be designed in the future (Solution 1) and how innovation will enable it (Solution 2) addressed, the next focus is how consumers will be involved in a recycling system. When I began my research in 2008, I was conflicted over the effectiveness of mandated refund-deposit programs for beverage containers, otherwise known as bottle bills; from one perspective, I saw it as a way to reduce waste, and from the other, I thought it was somewhat helpful, but inefficient, and perhaps harmful to the economy. After spending a few years researching the problem, I developed a set of important rules:

Rule 1: Due to the light weighting of aluminum cans and plastic bottles in markets, the secondary value of these materials is too low to be redeemed by the consumer.

Due to advances in material science, and razor thin profit margins in the beverage industry, the value of a used beverage container to a consumer is negligible. It is unworthy of their time to recycle in too many cases. Even the most valuable used beverage container to industry (the aluminum can) is worth less than 2 cents a container without a deposit. If these containers are economically worthless, there is little incentive to do anything other than chuck them in the trash can. Even if material design innovations makes these containers more valuable, it could still be worth less than the processing costs if the recycling system is inefficient.

Rule 2: Consumers in the United States will not tolerate higher prices for the sake of recycling strategies.

The U.S. beverage consumer products industry is a highly competitive/ price sensitive model, especially in the soft drinks sector. Coke and Pepsi are constantly waging a price war with the consumer for the smallest of profit margins; some bottled water companies can make as little as 3 cents for a case of plastic bottles! Furthermore, there have been no new refund/deposit laws enacted in the United States in the 21st century, with recent proposals in the 2014 election being voted down in various states. Despite the perception that Americans believe in sustainability programs, the findings are highly suggestive that consumers are unwilling to tolerate higher prices in order to help the environment. More expensive beverage containers (such as one-way and two-way glass bottles) are being phased out in place of cheaper market alternatives, such as PET plastic. Recycling advocacy groups have been unable to change public opinion in regard to this issue. Truth be told, consumers have prioritized low prices, then sustainability, if at all.

Rule 3: Recycling programs are (unintentionally) lowering the collection rate/market value of an efficient secondary packaging material (aluminum) while artificially increasing the collection rate/market value of inefficient packaging materials (plastic, glass, plastic-lined paper) in order to reduce externalities (e.g., landfilling and pollution) in a waste management strategy.

The data from my research has clearly found this to be the case. When recycling collection rates are prioritized over recycling reuse rates, these suboptimal conditions form, which is the basis of the recycling myth.

Due to collection being fostered as a waste management over an economic strategy, incremental innovation programs for recycling have proven to be unsuccessful in the United States and will never lead to the game-changing disruptive innovation that is required to truly address this packaging waste problem. Yes, in the United States, there are large differences in the recycling rates of voluntary recycling states (15–25%) versus mandated recycling states (70–75%). However, given the low value of recycled materials (Rule 1), and the high cost to process secondary materials, the solution must focus not only on higher collection rates but lower processing costs as well. Improving the material design of the container and its supply chain on the front-end will improve matters, but this will not solve the entire problem.

From my research, I have found that every current method of recycling collection had higher processing costs than the value of the materials themselves; curbside recycling, traditional refund/deposit system, and the California refund/deposit system all had costs higher than the value of the secondary material, even in some cases for aluminum! Even if efficient materials displaced inefficient ones, there is a potential for these gains to be lost through inefficient collection processes, thus requiring additional funding in order to maintain the value of the container. This is the tricky problem of collecting an item in the hundreds of millions, if not billions of low-cost items: ensuring the cost does not disrupt the consumers so that they will properly process items to be reused and not wasted. Paradoxically, the most efficient approach to processing beverage containers, at least from a processing cost standpoint, is through the landfilling of these materials by private mega-waste management companies, hermetically sealing them to rest forever in a grave of no reuse.

To improve the processing costs of secondary materials, why not focus on incentivizing good consumer behavior rather than punishing everyone with a container deposit? This makes sense: it is not economical to have a consumer who is already willing to recycle first to pay a deposit, which adds waste through transaction costs. A positive incentive for a consumer to recycle *without a deposit* will lead to higher recycling rates and lower processing costs for containers that aren't recycled. Taking out recycled containers without a transaction cost will improve the financial viability of the operation for those containers that aren't recycled. My research found this to be true: if given the choice and ability, most consumers will recycle a beverage container in order to avoid a 5- or 10-cent deposit. Therefore, deposits would only be charged to those unwilling to recycle, which would lead to a better funded refund-deposit system.

I have named this potential solution a *voluntary refund-deposit system* for beverage containers. As opposed to a mandated system that treats all consumers and materials as the same, and wastes transaction costs on consumers who are willing to recycle anyway, this system can account for differences in recycling behaviors through an information technology (IT) system that tracks the bottle, can, and cup from manufacturer to recycler to reuser, including consumer recycling behavior. Table 8.1 displays how this process can work.* It commences with a consumer purchasing beverages at a retail outlet. Through a loyalty card program (which already exists through most retailers), the consumer will be able to opt out of paying a deposit on the beverage container with a promise to return the containers within one month. If the consumer returns these containers to a sponsored drop-off site, the drop is recorded into the IT system, and no deposit will be charged. However, if the consumer does not drop off an equal number of containers during this period, the consumer's account will be automatically charged the deposit. Once the consumer returns the cans, the deposit is returned. If the consumer does

*For more details on the Voluntary Refund Deposit System, see Buffington, J., and Light Metals. 2014. TMS 2014 Annual Meeting and Exhibition. (January 01, 2014). "The Viability of a 'Voluntary Refund-Deposit System' for Aluminum Can Recycling in the U.S. *Tms Light Metals*, 913–918.

Table 8.1   Voluntary Refund Deposit vs. Traditional "Bottle Bill" Program Attributes

Attribute	'Bottle Bill' Program	Voluntary Refund-Deposit Program
Deposit on Container	Mandatory in 10 U.S. States	Only if consumer doesn't recycle
Self-Funded Program	Typically not in United States and Europe	Yes, by definition
National vs. State Program	State program (no uniformity)	National program (uniformity)
Innovation	Incremental	Disruptive (IT-based, tracking system)
Requirement	State referendum (none in 21st century)	No referendum; innovator and agreement in consumer products industry

Source: Buffington, 2013.

not opt out at retail, the deposit will be charged, and the refund-deposit system will be managed through the IT system.

In a research project I conducted, I calculated a comparison between the three collection models: states with no refund-deposit (R_1), states with mandated refund-deposit (R_2), and the proposed voluntary refund-deposit system (R_3). I calculated the starting recycling rate for the voluntary refund-deposit R_3 model as slightly lower than the mandated one (R_2), but the expectation is that, over time, a recycling culture will spread in non-refund deposit states, leading to rates of 70–75% or higher. Also, given that it is unlikely that non-deposit states will enact deposit legislation, a 66.5% recycling rate is much higher than the aggregate current rate of 40.6% across the entire United States.

The key to a voluntary refund-deposit system is its financial viability, leading to economic incentives for beverage can manufacturers and recycling entities without taxing producers and consumers. The current recycling rate for voluntary collection states (R_1 in my study) is 23%. The mandated R_2 model is underfunded in most states, being a heavy burden on state budgets, beverage manufacturers, and consumers who wish to recycle voluntarily; it also leads to high processing costs and does not lead to higher reuse rates for any material other than aluminum (which is undercollected). In summary, today's system is wrought with complexity, inefficiency, a lack of incentives, and, in many cases, fraud, both in the United States and

abroad. Yet despite these challenges, it has been the best program possible
for decades, at least until technology could achieve better results through
the voluntary refund-deposit system. With regard to the states today without a mandated program, it is clear that no new laws will be enacted in
the 39 voluntary states in the near future, if ever (Rule 2). In contrast, the
voluntary refund-deposit R_3 system has a starting collection rate increase
of over 25% in one year, which improves over time, is better funded than
the other two programs, and does not burden the individual who voluntarily recycles. For those who are against mandated recycling programs, just
recycle your cans and avoid the regulation! This is a compromise that the
beverage industry, state regulators, environmental advocates, and consumers can accept in order to balance the economy and environment.

As is shown in Table 8.2, this model enables companies to participate
in recycling as a market consideration, as opposed to being mandated
to do so. As we know from experience, organizations that choose to
participate in a market program are more motivated than ones that are
forced to participate.

Table 8.2   Comparison of Collection Models

	R_1 (Voluntary)	R_2 (Mandated)	R_3 (Hybrid)
Recycling Rate	23.0%	71.6%	66.5%
Oper. Cost/Can	1.91c	2.69c	0.9c
Surplus/(Deficit)/ Can	0.10c	1.60c	2.10c
Consumer	Not responsible	Responsible and burdened	Responsible, not burdened
Distributor/ Retailer	Not involved	Mandated	Choice
Redemption Center	Not Profitable	Not Profitable	Profitable, market-based
Curbside Program	Break-even, subsidize externalities	Break-even, greater subsidies for externalities	Market-based profitability, no subsidies
Government	Market-based, externalities not addressed	Externalities addressed, not market-based	Market-based and externalities addressed

Source: Buffington, 2013.

Similar to a mandated refund-deposit system, the voluntary refund-deposit system is funded through the unredeemed deposits of those who do not recycle but is administered more efficiently since processing costs are not incurred for UBC that would have been recycled anyway, a waste of processing costs. This improves the financials of the system since today's mandated recycling programs cannot break even if recycling rates exceed 70–80% because processing costs outweigh net unredeemed deposits. Therefore, there is a natural disincentive in today's bottle bill programs if recycling rates are too high, or a need to further tax the producers in order to account for the lack of economic value of the beverage containers being collected. When these programs are underfunded, the state must run a deficit, add a subsidy, take away from the consumer's deposit, or further tax the manufacturer/supplier. In contrast, my voluntary refund-deposit program lowers administration costs by excluding recycled containers that are voluntarily processed. In contrast to mandated programs where a disincentive exists for recycling, in my program, voluntary recycling is incentivized for the consumer, producer, and recycling program. Also, transaction cost efficiency is promulgated through information technology, as has been proven in other cases, such as the travel industry, financial services, and others. This technology will also improve the efficiency of the small scale, decentralized recycling industry, which currently cannot compete with the super-efficiency of today's primary material producers of petrochemicals and metals.

This program addresses the high cost of processing beverage containers outside of equilibrium, leading to a market-based model rather than a waste management mitigation strategy. In the next solution (Solution 4), I will address the issue of material value differences between aluminum, glass, plastic, and future materials. A voluntary refund-deposit system is a disruptive innovation in contrast to the mandated refund-deposit programs of incremental innovation as follows:

1) The value of a beverage container to the consumer will be based upon the choice that he and she makes; if the consumer promises to return and does, then there is no cost; however, if not, the consumer will be subject to the 5- or 10-cent deposit. This would ensure that a non-return leads to a can on the street having at least a 5cent value, ensuring an economically viable supply chain function.

2) If the consumers cannot tolerate paying a deposit on a case of cans or bottles ($1.20–$2.40 a case), they can opt out of the program by promising to return the cans in a month's period. Because this will become the consumer's choice, it does not unfairly punish a consumer with a deposit mandate nor allow the

individual to avoid the responsibility for recycling. The beverage manufac-turing supply chain should also be in favor since deposits are not mandated; advocacy groups in favor of "bottle bills" should consider this to be a practi-cal compromise, given the lack of history in the United States for favoring legislation.

3) Distributors and retailers who do not want to participate in recycling pro-grams will not be required to do so. Today, their involvement is often man-dated, and in many cases, it can lead to their losing money or focus on their core business of selling beverages. Since the voluntary refund-deposit pro-gram will be self-funded, by definition, there will be no need to mandate participation, as efficient vendors that process UBC will be able to make a profit.

4) Government municipalities will not be required to participate; therefore, they will not be subject to program deficits, nor, for that matter, will they take advantage of program surpluses. This system will not require any gov-ernment involvement to implement. Government involvement should be restricted to overseeing the program to ensure it enables both economic growth and environmental sustainability.

5) The program will lead to higher aluminum recycling rates, enabling alumi-num can fabricators to enjoy a larger secondary supply of material, effectively bringing costs down for the entire supply chain, including consumers. Higher collection rates will exist for other materials as well, as will be addressed in Solution 4.

The goal of this program is to achieve the highest recycling rates pos-sible at the most efficient processing cost and with the least burden on all stakeholders. Through IT system innovation, there is no reason why UBC cannot be tracked in this manner and rationalized. It also brings a healthy compromise between those who are against a deposit system and those in favor of it. The use of a voluntary refund-deposit system would also have the potential to reduce the level of damage and contam-ination in less recycled materials, such as glass and plastic, and therefore, through market means, increase its content reuse in beverage contain-ers while having a positive impact on the financials for the beverage manufacturers.

SOLUTION 4: MARKET BASED EXTENDED PRODUCER RESPONSIBILITY (M-EPR)

Even if Solutions 1–3 are achieved, beverage manufacturers and packaging material suppliers must be held directly accountable for the sustainability of its materials but in a manner that enables change. History has often proven that when faced with disruptive innovation, dominant market players resist change, so there is a possibility that conventional dominant design materials, like today's PET plastic, could stay in the marketplace even after disruptive innovation, due to lower costs. Solution 3 (voluntary refund-deposit system), led by disruptive technology and process, places the onus on the consumer to recycle but does not address the problem of how to handle materials with different reuse rates and financial values in the open market (aluminum, plastic, and glass).

In the United States, there are 69 extended producer responsibility programs in place for various products, such as paints, tires, batteries, and electronics, but to an insignificant degree for packaging materials.[1] In contrast, many European nations have had some form of mandatory EPR program in place for decades, but it has led only to incremental improvements, at best, in regard to material use and reuse. In some European countries, there is a Green Dot System, where the producer must make a financial contribution toward the recycling of that beverage container. In Sweden, Returpack and FTI are responsible for implementing this program, with funding from the beverage companies. The intent of these programs in Europe is more to mitigate waste than to address the market value of secondary materials. There are examples of EPR programs, such as in Ontario, Canada, where the cost to the producer is calculated based upon use and the value of the material that can lead to a credit balance for products such as aluminum, yet in its Waste Reduction Act, producers are responsible for 100% of the cost of the EPR program even though they cannot control how materials are collected or processed and what materials are collected! Other well-regarded programs such as the Green Dot System in Germany are criticized for being a privatized system that put their own financial interests first, at the price of effective recycling rates and processing efficiency. As a result, there is a movement within Germany to deprivatize the system and call for higher recycling quotas. As is shown in chapter six, bottle-to-bottle recycling rates for PET plastic in these European nations is not much different than in the United States, which suggests that these EPR programs are not working.

Presently, there are no significant EPR programs in the United States for beverage packaging materials, but there are special interest groups pushing for their enactment. Studies have been conducted in the United States by nonprofit organizations, such as Recycling Reinvented, and found that an EPR system can be implemented that will increase recycling rates and keep costs flat to the current system. However, they mention that costs would have to be embedded into packaging materials in order to finance the system, passing the excess costs on to the consumer.[2] Nestlé Waters, North America (NWMA) is perhaps the most visible beverage manufacturer in the United States officially promoting the EPR concept, with past CEO Kim Jeffery stating that the "system is broken," while other companies in the industry did not follow suit.[3] Targeting a plastic bottle recycling rate of 60%, over twice the current rate, Nestlé (a European-based company) communicates its desire for a voluntary *take back* program. Nestlé faces resistance from its competitors and would create a market imbalance if it went off on its own. Yet the data shows that EPR programs are not proven to have a significant impact on incentives and disincentives regarding well-designed (e.g., aluminum) and poorly designed (e.g., PET) materials, respectively. Therefore, just as in Solution 3, transaction costs are wasted on consumers who would have recycled voluntarily. The same is true for Solution 4 and the inefficiency of wasting costs on valuable materials, such as aluminum, similarly to less valuable ones, such as PET.

Conceptually, EPR and bottle bill programs are the right idea: to hold producers and consumers accountable for the cost of collecting and processing materials, minus the value of the material. But just as today's bottle bill programs have done little to change the nature of beverage materials used by consumers, EPR programs have not led to improvements in material science. If implemented effectively, these programs should lead to a change in material types used, not a focus on the reduction of landfilling. To address this problem, I have developed the concept of a market-based extended producer responsibility (M-EPR) program that requires the producer to be responsible for the reuse rates of its packaging material. My proposed solution is the use of a voluntary M-EPR system that is not to mandate producer responsibility on all beverage packaging equally but rather to make it market-based upon the environmental viability of the material *for same or higher reuse or biodegradation*. For example, if a commodity like aluminum is recycled and reused (for the same or higher use) at the targeted rate, then no regulatory mandate should be in place. In contrast, if a plastic or glass commodity is below the targeted rate, and/or downcycled (e.g., lesser uses, incineration, sent to developing countries), the beverage manufacturer will be required to increase its reuse rate, pay penalties, or change the material. The goal of the market-based program is a can-to-can or bottle-to-bottle reuse rate. This is change from a conventional EPR program that equally apportions costs and reuses materials for lesser purposes, which doesn't lead to improvement.

The intention of an M-EPR program, in contrast to a conventional EPR, is to hold the producer responsible for the design of the product, or, as the industry likes to say, design for recycle (DFR). If an EPR program is not leading to changes in material design that increase reuse, it is an ineffective program regardless of recycling rates. How is a manufacturer in Europe really being held accountable for the low plastic reuse rates as discussed in chapter six? An effective M-EPR program will incentivize an increase in the reuse rate of the lowest packaging material closer to the material with the highest percentage if they are used for the same function. For example, if the starting point reuse target for all materials is 50%, the goal would be immediately achieved for aluminum cans where the secondary market demand is already above 50%. In contrast, the PET plastic bottle would be significantly below this targeted rate at 12–18%; however, through Solution 3, there is a potential for higher collection rates. Note that under this definition, Europe's high recycle/reuse rates and self-professed zero waste culture would be viewed under a much tighter definition of sustainability! Under a market-based approach, the PET producer would immediately need to think of its short-term and long-term strategy in regards to the reuse of its product, which still isn't happening in some of the top sustainability countries today. It is likely that the company would not be able to quickly cover the gap between 12–18% and 50% too soon and would be subject to dumping fees, perhaps after a transition period. The company would pass on these additional costs to the consumer, which would lead to the following actions:

1) Soft drinks in PET plastic and glass would increase in cost in comparison to aluminum cans. This is inevitable and equitable given the low reuse rate of these materials. Materials not being recycled today, such as coffee cups, convenience store/quick-service cups, K-cups, and sippy pouches would increase in cost as well.

2) Price-sensitive consumers would start choosing more environmentally friendly packaging, making less environmentally friendly packaging not market viable. The gap between what is optimized from an environmental standpoint and from an economic standpoint would begin to close.

3) Producer companies would either redesign these poorly reused materials to achieve 50% reuse or develop alternatives. Innovation would be critical to maintaining its markets!

Soft drink and beer companies would be expected to migrate to a higher mix of aluminum from plastic/glass due to consumer demand changes. Currently, plastic has the lowest supply chain costs (the reason for market growth), but this would change, given the added M-EPR fees, at least until a disruptive innovation occurs in regard to the material design of PET or its alternative. As such, bioplastics would not be viewed as sustainable just because they are organic if they had the same reuse rate as petroleum-based plastic. Single-use glass packaging would become cost prohibitive and would either be discontinued altogether or restricted to niche packaging where the consumer is willing to pay a premium. Given the thermal requirements for coffee, perhaps a bio-degradability goal (rather than recyclability) could be used for coffee cups. It is interesting that these cups are outside of the present recycling programs despite the size of these materials in the trash waste stream. In this new system, plastic-lined paper cups would be subject to recycling, as is the case with aluminum, glass, plastic, and less traditionally recycled packaging forms, such as wine/liquor bottles, K-cups, aluminum foiled pouches, and plastic convenience store/QSR cups; it is puzzling to me how these nontraditional containers are allowed to exist outside of nearly all of today's mandatory recycling programs.

The same IT system innovation developed for the voluntary refund-deposit system would be used for the market-based EPR. The percent of recycled content in a packaging container and its volume sales would be periodically audited to determine the M-EPR fees associated with the packaging material. Also included could be an incentive system implemented for packages that exceed the targeted secondary goal. The issue of decomposable packaging will have to be addressed as well. Information technology has the potential today to disruptively change the recycling industry for the better! By using technology to track real innovation rather than greenmailing programs, change will occur.

The end result of Solution 4 is to hold the producer responsible for the design of the packaging material but with a market-based approach that provides control to the beverage manufacturer. Unlike conventional EPR programs, the emphasis and focus for this solution is *like for like* or better reuse content through improved design and manufacturing, not simply higher recycling rates in a waste management strategy that includes methods such as incineration or even landfilling. Much like the voluntary refund-deposit system with the consumer, the producer will be responsible for regulation only if targets are not achieved; a consumer is subject only if he or she doesn't recycle, and a producer is subject only if its packaging does not meet required reuse rates. In the United States, there is a strong resistance to traditional EPR programs in the consumer beverage industry for good reason. Making this program market-based enables the producers to control their own destiny rather than be subject to bureaucratic mandates. As a result, beverage and package manufacturers would feel empowered to address the problem rather than pay for a program where it has little or no control. This program would be good news for the aluminum can industry and innovators developing new solutions, pose challenges to status quo material manufacturers (PET, glass, plastic-lined paper) who are focused on low front-end costs and incremental innovation, and jump-start disruptive innovation entrepreneurs! It would also close the loop by including more beverage containers types. Consumers and producers would be incentivized to make proper environmental choices, and innovators would be encouraged to develop the next generation of beverage packaging. Since most consumers are resistant to price increases, manufacturers would be incentivized to offer packaging material that does not add costs due to low reuse rates. The municipality will generate higher dumping fees that are related to the reusability of the packaging material and likely dump fewer containers that actually have market value. It is also possible that reusable glass bottles could make a return, at least in limited markets.

An M-EPR program balances economics and the environment through a jumpstart of the innovation market, the improvement of the reusability of beverage containers, and a reduction in landfilling—all without financially penalizing beverage manufacturers who are seeking legitimate material science improvements. Most importantly, an M-EPR could potentially be the impetus to drive the disruptive innovation engine that is required to solve our massive packaging waste problem!

SOLUTION 5: REDUCE WASTE, NOT USE

To summarize, in Solution 1, Material Scientist 3.0 takes over for the human designer Material Scientist 2.0 in order to achieve transformative designs to mirror nature's process. Solution 2 leads to a new approach of innovation in which the entrepreneurial process is led by crowdsourcing, prize inducement competition, and supply chain transformation to create good or no packaging. Solution 3 changes the economic equation of consumer collection through a deposit opt-out program for those who voluntarily recycle. Finally, Solution 4 begins to transform the materials used in mass markets by incentivizing high reuse/biodegradable materials and penalizing bad performance. What is left for the United States and other countries to achieve a packaging revolution? Simple: consumers should be focused on reducing waste in order to improve both the economy and the environment; a new culture must be established.

Americans believe strongly in the rights of the individual, including the right for producers to sell products to consumers, even to the point of overconsumption. Today, the recycling movement has sometimes crossed the line, casting judgement on consumption rather than waste. As a result, the main argument pits the economy against the environment rather than optimizing both. In this book, I am advocating policy that enables both economic growth and environmental sustainability. Throughout the history of our planet, evolutionary progress has been achieved by more, not less. A model of consumer austerity is not only un-American, and anti-evolution, but also unnecessary. Peter Diamandis, the founder of the X Prize, is perhaps the most articulate innovator on this topic, encouraging change in our world through progress and consumption, and not an end-to-growth mantra that is repeated too often by economists and environmentalists. Of course, excessive use can be viewed as a form of waste, so I am not suggesting that this is in any way a good solution; rather, I seek practical, sustainable growth.

After World War II, the narrative in the United States was that American production would need to spew out a bunch of consumer articles continuously in order to maintain economic growth. To drive growth, the concepts of planned obsolescence, and then the throwaway society, arose, in which packaging waste was a large part—up to a third of all consumer waste. Marketers and strategists appeared to believe that growing markets through waste was necessary since the average American already had everything he and she needed. Eventually, this would become an unbelievably flawed and dangerous model that led to the analogy of being on a tiger we couldn't get off of without being eaten. It's a model that will inevitably lead to failure, like a Ponzi scheme, and we are beginning to see the short-sightedness of the strategy. This is an irresponsible approach to consumerism that has led us to the recycling myth that exists today. This is an inevitable problem that we, or our children, will face in our lifetimes.

Solutions 1–4 will be the easiest to accomplish in America because they can be driven through innovation, technology, and some policy change. Solution 5 will be much more difficult, as it is related to culture—how we live our lives. What would be the point of improving the relationships between industry and our environment without the consumer driving change? I'm optimistic that this can occur, but an improvement in the educational systems and moral values must be fostered alongside the growth in technology for the best effect. Waste is not just sin against the environment that can be magically resolved with science, technology, and innovation; aversion to it must become a part of our culture again as well. In the end, if we respect the value of resources, we serve others and ourselves as well, both in the present and for the future.

Chapter 9

2020: The Happy Cup Reality

2020: THE END TO THE RECYCLING MYTH

Are the solutions from this book just pie in the sky concepts or real possibilities? Would a nation, state, local municipality, or even a company within the supply chain give these disruptive innovation solutions an opportunity as an alternative to doing nothing or to mandated or voluntary recycling programs? For this to happen, someone will need to catalyze change: a beverage manufacturer; a recycling advocacy nonprofit organization; a legislator; or even someone from outside of the current system, such as an entrepreneur or philanthropist with a keen interest in the environment and economy. This catalyst will foster change, likely not just for the improvement of the environment but for a profit as well, which is often the driver of disruptive innovation.

Could it be an industry insider who drive change, as Bill Coors did over fifty years ago with the aluminum can and Cash for Cans recycling program? Research literature shows that this is unlikely; it is often incremental change that is driven by the dominant players in the market while radical change is driven by entrepreneurs. This is because the existing market players are very focused on the requirements of the present customer, and the customers cannot envision grand change. As was noted in chapter four, the U.S. consumer often appears to be content with a happy cup fallacy approach to marketing, low prices, and superficial sustainability programs. In regard to creating change by listening to the customer, Steve Jobs noted, "It's really hard to design products by focus groups. A lot of times, people don't know what they want until you show it to them."[1] After 40 years of recycling programs, it is clear that disruptive change is unlikely to be driven by consumers, producers, advocates, or governments. Just as was the case with the iPhone in 2007, innovation must come from the outside to transform packaging in the consumer beverage industry.

The disruptive innovation approach to the environment and economy could begin with an entrepreneurial catalyst through the development of a Voluntary Refund-Deposit Program. An entrepreneur could develop the software system to implement at retail locations within its point of sale kiosks and loyalty reward programs, along with developing a collection Reverse Vending Machine, or EarthVM, as I illustrated in my research. Upon development, the entrepreneur can deploy the solution across an entire municipality with the software/hardware solution or maybe even across a beverage manufacturer or retailer, introducing this as a cornerstone of its sustainability program to kick-start change, much like Bill Coors's Cash for Cans program, which was introduced despite opposition from its counterparts. Walmart is a candidate for implementing the program at retail outlets, driving change through its market position. In short, change can be introduced at a micro-level or within an entire jurisdiction through the entrepreneur's system.

What's important for this entrepreneurial catalyst or early adopter is a recycling and reuse program driven by more efficient processes, superior technology, and collaboration across the supply chain. Compromise is the key to achieving these results within the supply chain: environmental advocates must let go of the bottle bill program as the only way to solve the problem, joining forces with consumer product companies. Or a beverage manufacturer could be viewed as a leader in this voluntary process, leading to some of the eleven bottle bill states, testing these disruptive innovations as an alternative to their existing programs. Through compromise between bottle bill and non-bottle bill states, the beverage manufacturers, distributors, retailers, and recycling/waste management companies form a consortium among the consumer beverage industry and retail vendors to direct the work effort to build compatibility with this technology system in its manufacturing plants, distribution/retail outlets, and back-end waste management and recycling operations. Financial resources previously spent on happy-cup campaigns can be reallocated to this consortium, leading to no additional costs and showing the consumer that recycling and reuse will no longer be considered only a marketing strategy but also an environmental responsibility. Environmental advocates and government legislators may finally see sincere efforts for change in industry sustainability programs.

With the consortium in place, further development to the Voluntary Refund-Deposit and M-EPR is formalized, taking two years to fully develop, with a nationwide kick-off in 2018 and full deployment by 2020. The system will also be an open source innovation platform to enable crowdsourcing and prize inducement competition in order to change the innovation paradigm and make it happen sooner. Government involvement is encouraged in the process as well as support from environmental advocates in order to establish recycling, reuse, and biodegradation standards. Experts from the pro-recycling advocacy groups, such as CRI, should contribute their functional expertise.

In the first year of the rollout, system and process issues need to be addressed, as is the case in any large rollout. Also, changes in behavior are required of all parties in the supply chain. Recycling rates are naturally higher in the 11 ex-bottle bill states than in the 39 voluntary states, yet recycling rates, in aggregate, will increase from less than 40% to 50% in one year, the highest one-year growth since the onset of these programs! By 2020, the targeted rate of 60% aggregate will be achieved, due to adoption by former non-recyclers; efficiency across state lines due to uniformity; tracking via the IT system; and the fact that all stakeholders are working together, rather than against each other, in order to solve the problem. Under public pressure, a few states that cling to their bottle bill laws will join this voluntary, self-funded program by 2022, leading to even higher results due to one uniform system of recycling across all states and beverage containers in the United States.

After the successful rollout of the Voluntary Refund-Deposit system is the kick-off of the Market-Based Extended Producer Responsibility program, or M-EPR, based upon the same technology platform. The adoption of this program is more contentious, pitting beverage companies against one another, with some favoring a market-based approach to producer responsibility while others are against it—whether it's voluntary or mandated. Even within the same market sector, such as bottled water, there are companies in favor of an M-EPR, like Nestlé Waters, and others dead set against it, given a market strategy position of offering low-priced water in low-priced, nonrecyclable plastic beverage containers.

The challenge becomes not a competition between those in favor or against, but rather between beverage manufacturer innovators who seek disruptive change and those who present only faux environmental campaigns. In the 1960s through the 2000s, beverage innovation was driven by making packaging lightweight, portable, and disposable, but the new innovators will offer concepts such as 3D packaging, edible/

biodegradable packaging, or even no packaging at all, among other ideas. As a result of the pressures from the upcoming packaging revolution, the M-EPR moves forward. In the first year, fines are not levied for aluminum cans but are so for most beverage containers, such as PET plastic bottles, glass bottles, plastic-lined paper cups, and plastic take-away cups.

For some bottled water companies that fight the M-EPR model, sales begin to fall, as they gain bad publicity in comparison to other companies that disclose their plans to address the problem. In many cases, prices rise for beverage manufacturers who are levied fines, leading to some companies ramping up their efforts to design reusable/biodegradable bottles. In contrast, other companies simply balk at the concept of an M-EPR and lobby to repeal these efforts. Consumers respond by using less of the higher-priced products, by supporting companies (even with temporarily higher prices) that are sincere in their efforts to design new containers, or by joining the fight for low-cost, environmentally friendly packaging. Companies that promote market-based competitions are also viewed favorably by consumers and perceived as true change agents.

Coffeehouses, quick service restaurants (QSRs), and convenience stores also struggle at the onset of the M-EPR, given that they didn't participate in conventional recycling programs previously. Even accounting for a lower starting reuse target rate of 30% instead of the standard 50% for containers within existing bottle bill programs, they miss the mark by significant amounts, leading to fines levied on their containers. Since many of these retailers cater to low-cost markets, passing on the higher costs to consumers is seen as untenable, and many of them petition for extensions or even an outright appeal. However, other retailers become more creative, establishing reusable to-go cup programs in order to avoid the fines. In contrast to the low margin, high volume QSRs and convenience stores, higher-end coffeehouses and other specialty shops find the greater container challenge to be involving their customers in the process. At first the message is that "costs will be higher, but these higher costs will lead to a more sustainable packaging material." Consumers accept this basis as a matter of principle. Eventually, innovative retailers become the leaders in 3D printing of sustainable beverage containers, leading to higher sales. In contrast to other consumer beverage providers, these retailers find the M-EPR to be an opportunity to attract greater market share through real sustainability applications. In this new model, companies that are transparent about their container design misgivings and their efforts to redesign are appreciated and supported in contrast to conventional wisdom that happy-cup campaigns must be implemented to fool the customer.

By 2025, the unthinkable happens: given the use of the Voluntary Refund-Deposit System and M-EPR, and the maturity of supercomputers and 3D printing, radical innovations begin to replace plastic and glass bottles, plastic-lined cups, and plastic cups! In 2016, crowdsourcing appeals are announced, leading to disruptive market innovations from out of the blue, and, nine years later, true market innovation with scale to provide at least 50% of the aggregate existing beverage market emerges. The costs of these materials start off slightly higher, but some rebates are provided from the M-EPR to support innovation. Eventually, the costs are lower than the conventional material; therefore, by 2027, there is no conventional (plasticized) plastic and glass used in the beverage packaging industry. Aluminum costs continue to fall through its high reuse, and these new materials lead to a beverage supply chain system of zero waste, economic growth, profitability, and the industrial supply chain being extended to nature.

At the beginning of this process, 1 billion cans, bottles, and cups used each day in the United States were recycled approximately 25% of the time (with coffee cups, QSR, and convenience store cups included). More than 700 million containers head to the landfill on a daily basis. In the future model of disruptive innovation, the use of supercomputers, and policy focus, recycling rates progress to 60% in two years and 75% after five years, with reuse rates increasing to 50% and higher. After ten years, this program will be a world leader for economic and environmental balance: 1% waste (almost zero waste without incineration), 90% recycling, 75% reuse, significant biodegradation, and a financially viable program that enables consumer growth at American standards. This is truly a disruptive innovation model!

Transformation can really occur within a decade if an entrepreneurial catalyst is introduced and embraced, driving alignment and collaboration among all stakeholders rather than emphasizing ideological divides. In this process, not only are new economic market opportunities made possible, but the environment is improved, as opposed to becoming slightly less worse. Perhaps in the longer term, advocacy groups like the 5 Gyres will be able to clean the ocean of synthetic plastic with confidence that future beverage container litter will be biodegradable and, therefore, actually good for the ocean! At first it seems strange to throw a bottle on the ground in the forest during a walk, but after seeing deer and rabbits safely grazing upon it, you realize that your portable refreshment at one moment becomes part of the natural cycle of life in the next. Is this level of symbiosis between nature and our consumer products industry possible? Why not? Would anyone have thought 50 years ago that Americans would be using 1 billion beverage containers in a single day? Perhaps the greatest struggle in achieving all of this is our lack of imagination.

WHY CHANGE?

In this book, I've developed an alternative to the current Option 0 of *do nothing* and let the waste pile into landfills and Option 1 of imposing nationwide mandates that collect more than is reused, reduce consumption, and burn containers under clever waste-to-energy zero-waste claims. But is this really an issue of focus for the American public, or of any nation, given some of the other problems in the news? On the surface, perhaps not, but if reducing packaging waste leads to economic growth, it should be a policy focus. Achieving a program that enables economic growth while improving the environment should be on everyone's agenda.

Not until America, or any nation for that matter, considers what's good for the economy to be good for the environment, and vice versa, will this problem be addressed—a theme that, of course, goes beyond just beverages. With the use of supercomputers to optimally design beverage containers and its supply chains extending to nature, the use of novel materials unknown before, the use of information technology advancements to track and manage the secondary markets within a collaborative manner, a consumer's appetite for real change, and the unleashing of new forms of innovation to solve for it, this waste problem can actually be solved rather than just mitigated. All that's left is our own personal attitudes and biases, and how we live our lives in general. As soon as we are willing to change, we can change the paradigm for this and other conventional problems. Yet this is the greatest challenge we face: whether we are ready to address such problems head on, working together on the *ends* rather than the *means*. Or will our apathy continue to lead to greater waste? To solve a problem of this magnitude, we must not get recycling program deposit systems "in our blood," but rather disruptive innovations "in our souls." Once we do, we will begin to understand how our old culture of waste was not just a poor way to treat nature but a lack of respect for ourselves and our future as well. Let's get on with it!

Notes

CHAPTER 1: THE MYTH

1. LaCapra, Veronique. 2015. "With 'Single-Stream' Recycling, Convenience Comes At a Cost." NPR.org (March 31): http://www.npr.org /2015/03/31/396319000/with-single-stream-recycling-convenience-comes-at-a -cost.

2. Pierce, Lisa McTique. 2012. "North American Packaging Market Remains Wary about the Future." *Packaging Digest* (November 11): http:// www.packagingdigest.com/decorative-materials/north-american-packaging -market-remains-wary-about-future.

3. LaCapra, Veronique. 2015. "With 'Single-Stream' Recycling, Convenience Comes At a Cost." NPR.org (March 31): http://www.npr.org /2015/ 03/31/396319000/with-single-stream-recycling-convenience-comes-at-a -cost.

4. Gitlitz, Jenny. 2013. "Bottled Up." Container Recycling Institute. http:// www.container-recycling.org/index.php/publications/2013-bottled-up-report.

5. Kolbert, Elizabeth. 2009. "Hosed." *The New Yorker.* (November 16): http://www.newyorker.com/magazine/2009/11/16/hosed.

6. Pasternack, Alex. 2013. "The Most Watched Load of Garbage in the Memory of Man." *Motherboard.* (May 13): http://motherboard.vice.com/blog /the-mobro-4000.

7. Ibid.

8. Pellow, David. 2002. *Garbage Wars: The Struggle for Environmental Justice in Chicago.* Cambridge, Mass: MIT Press.

9. Gutis, Philip. 1987. "For Alabamian, L.I.'S Garbage Is Dream Gone Bad." *The New York Times.* (May 6): http://www.nytimes.com/1987/05/06/nyregion /for-alabamian-li-s-garbage-is-dream-gone-bad.html.

10. Pasternack, Alex. 2012. "An Unfinished Park Built Atop a Giant Pile of Trash Is Making New York City Millions." *Motherboard*. (June 26): http://motherboard.vice.com/blog/how-an-unfinished-park-built-atop-a-giant-pile-of-trash-is-making-new-york-city-millions.

11. HenryFund.org "Waste Management Industry." http://tippie.uiowa.edu/henry/reports14/waste_collection.pdf.

12. "Waste Market Overview and Outlook." *Waste Management Business Journal*. 2012. http://www.wastebusinessjournal.com/overview.htm.

13. Davis, Aaron. 2015. "American Recycling Is Stalling, and the Big Blue Bin Is One Reason Why." *Washington Post*. (June 20) http://www.washingtonpost.com/local/dc-politics/american-recycling-is-stalling-and-the-big-blue-bin-is-one-reason-why/2015/06/20/914735e4-1610-11e5-9ddc-e3353542100c_story.html.

14. "Municipal Solid Waste." U.S. Environmental Protection Agency. http://www.epa.gov/epawaste/nonhaz/municipal/.

15. Ibid.

16. Rossignol, Kevin. 2014. "The Ways We Waste: The Long Road to Sustainable Waste Management." *EPA Online*. (August 21): http://eponline.com/articles/2014/08/21/the-ways-we-waste.aspx.

17. Davis, Aaron. 2015. "American Recycling Is Stalling, and the Big Blue Bin Is One Reason Why." *Washington Post*. (June 20): http://www.washingtonpost.com/local/dc-politics/american-recycling-is-stalling-and-the-big-blue-bin-is-one-reason-why/2015/06/20/914735e4-1610-11e5-9ddc-e3353542100c_story.html.

18. Makower, Joel. 2013. "Exit Interview: Kim Jeffery, Nestle Water." *GreenBiz.com*. (March 4): http://www.greenbiz.com/blog/2013/03/04/exit-interview-kim-jeffrey-nestle-water.

CHAPTER 2: FRANKENSTEIN'S BOTTLE: A PROBLEM OF MATH

1. Makower, Joel. 2014. "Why the World's Biggest Companies Are Investing in Recycling." *GreenBiz.com*. (August 18): http://www.greenbiz.com/blog/2014/08/18/why-worlds-biggest-companies-are-investing-recycling.

2. Ibid.

3. Putrich, Gayle. 2015. "Study: Plastic Ocean Waste on the Rise." *Plastic News*. (February 12): http://www.plasticsnews.com/article/20150212/NEWS/150219956/study-plastic-ocean-waste-on-the-rise.

4. Obbard, R., Sadri, Y., Wong, A., Khitun, I., Baker, and Thompson, R. C. 2014. "Global Warming Releases Microplastic Legacy Frozen in Arctic Sea Ice." *Earth's Future*, 2, 315–320, doi:10.1002/2014EF000240.

5. U.S. Geological Survey. 2015. Mineral Commodity Summaries. (January).

6. de Huib, V., Purchase, R., de Groot, H., and Lankhorst, R. 2013. *Solar Fuels and Artificial Photosynthesis Science and Innovation to Change Our Future Energy Options*. Wageningen: BioSolar Cells.

7. Berner, R. A. 2004. *The Phanerozoic Carbon Cycle: CO2 and O2*. Oxford University Press.

8. Wrigley, E. A. 2013. "Energy and the English Revolution." Philosophical Transactions A. (January 28): http://rsta.royalsocietypublishing.org/content /371/1986/20110568.figures-only.

9. Landes, David S. 1969. *The Unbound Prometheus: Technological Change and Industrial Development in Western Europe from 1750 to the Present*. London: Cambridge U.P.

10. Wrigley, Tony. 2011. "Opening Pandora's Box: A New Look at the Industrial Revolution." *VOX CEPR's Policy Portal*. (July 22): http://www.voxeu.org /article/industrial-revolution-energy-revolution.

11. U.K. Department of Energy and Climate Change. 2014. *Estimated Average Calorific Value of Fuels*. https://www.gov.uk/government/statistics/dukes -calorific-values.

12. Wrigley, Tony. 2011. "Opening Pandora's Box: A New Look at the Industrial Revolution." *VOX CEPR's Policy Portal*. (July 22): http://www.voxeu.org /article/industrial-revolution-energy-revolution.

13. Hagens, Nate. 2014. "Humans and Earth: Transitioning from Teenagers to Adults as a Species (lecture)." (May 15): http://www.themonkeytrap.us /humans-and-earth-transitioning-from-teenagers-to-adults-as-a-species-lecture.

14. Stern, Roger. 2012. "Oil Scarcity Ideology in US National Security Policy, 1909–1980." Princeton University. http://www.princeton.edu/oeme /papers/Roger%20Stern%20Oil%20Scarcity%20Ideology%20in%20US%20 National%20Security%20Policy.

15. Schaber, Stephen. 2012. "Why Napoleon Offered a Prize for Inventing Canned Food." *NPR Planet Money*. (March 5): http://www.npr.org/blogs/money /2012/03/01/147751097/why-napoleon-offered-a-prize-for-inventing-canned-food.

16. U.S. Department of the Interior. *Historic Glass Bottle Identification & Information Website*. http://www.sha.org/bottle/.

17. MacKerron, C. 2015. "Waste and Opportunity 2015: Environmental Progress and Challenges in Food, Beverage, and Consumer Goods Packaging." *As You Sow* (January).

18. Ibid.

19. Clarke, Chris. 2014. "When Recycling Becomes a Dirty Business." *takepart.com*. (August 27): http://www.takepart.com/article/2014/08/26/when -recycling-gets-messy.

20. Standen, Amy. 2014. "Should Wine Bottles Carry a Deposit?" *KQED Science*. (February 3: http://blogs.kqed.org/science/audio/should-wine-bottles -carry-a-deposit/.

21. History.com. 1935: "First Canned Beer Goes on Sale." http://www.history .com/this-day-in-history/first-canned-beer-goes-on-sale.

22. Henion, Karl E. II. 1981. "Energy Usage and the Conserver Society: Review of the 1979 AMA Conference on Ecological Marketing." *Journal of Consumer Research*. 8 (3): 339.

23. Kean, Sam. 2010. "Blogging the Periodic Table." *Slate.com*. (July 30): http://www.slate.com/articles/health_and_science/elements/features/2010/blogging_the_periodic_table/aluminum_it_used_to_be_more_precious_than_gold.html.

24. Ibid.

25. Canby, T. Y. 1978. "Aluminum, the Magic Metal." *National Geographic*. (August), p. 204.

26. Buffington, J. and Peterson, R. 2013. "Defining a Closed-Loop U.S. Aluminum Can Supply Chain Through Technical Design and Supply Chain Innovation." *JOM*, Volume 8 (June), pp. 941–950.

27. Elliott, B. 2014. "Aluminum Can Recycling Rate Flat in 2013." *Resource Recycling*. (October 1): http://resource-recycling.com/node/5321.

28. Ibid.

29. Freinkel, S. 2011. "Plastic: A Toxic Love Story." New York: Houghton Mifflin Harcourt.

30. Esposito, F. 2015. "PET Bottle Resin, Recycled PE Prices Drop." *Plastics News*. (February 12): http://www.plasticsnews.com/article/20150212/NEWS/150219957/pet-bottle-resin-recycled-pe-prices-drop.

31. PET Resin Association. "Little-Known Facts about PET Plastic." http://www.petresin.org/news_didyouknow.asp

32. MacKerron, C. 2015. "Waste and Opportunity 2015: Environmental Progress and Challenges in Food, Beverage, and Consumer Goods Packaging." *As You Sow*. (January).

33. Ibid.

34. *Resource Recycling*. 2014. "EU Pushes for More PET Recycling." (July 16): http://resource-recycling.com/node/5083.

35. MacKerron, C. 2015. "Waste and Opportunity 2015: Environmental Progress and Challenges in Food, Beverage, and Consumer Goods Packaging." *As You Sow*. (January).

36. Ibid.

37. The Electronic Law Library. 1995. "The Actual Facts about the McDonald's Coffee Case." http://www.lectlaw.com/files/cur78.htm.

38. Kamenetz, Anya. 2010. "The Starbucks Cup Dilemma." *Fast Company*. (October 20): http://www.fastcompany.com/1693703/starbucks-cup-dilemma.

39. Brown, Elizabeth. 2015. Cop Suing Starbucks Over Spilled Coffee. *Reason*. http://reason.com/blog/2015/05/05/cop-suing-starbucks-over-spilled-coffee.

40. Hamblin, James. 2015. "A Brewing Problem." *The Atlantic*. (March 2): http://www.theatlantic.com/technology/archive/2015/03/the-abominable-k-cup-coffee-pod-environment-problem/386501/.

41. Godoy, Maria. 2015. "Coffee Horror: Parody Pokes at Environmental Absurdity of K-Cups." *NPR.org*. (January 28): http://www.npr.org/blogs/thesalt/2015/01/28/379395819/coffee-horror-parody-pokes-at-environmental-absurdity-of-k-cups.

42. MacKerron, C. 2015. "Waste and Opportunity 2015: Environmental Progress and Challenges in Food, Beverage, and Consumer Goods Packaging." *As You Sow*. (January).

43. Collins, S. 2013. "Bottled Up: (2000–2010) Beverage Container Recycling Stagnates." *Container Recycling Institute*. http://www.container-recycling.org/index.php/publications.

44. Savinov, V., & Smithers Pira (Firm). (2014). *The Future of Global PET Packaging to 2019*.

CHAPTER 3: THE THROWAWAY SUPPLY CHAIN

1. Sizelove, R. 2011. "Nine in Ten Adults Recycle, but Only Half Do So Daily." *Ipsos*. (July 13): http://www.ipsos-na.com/news-polls/pressrelease.aspx?id=5285.

2. Packard, Vance. 1960. *The Waste Makers*. New York: D. McKay Co.

3. Jensen, Richard. 2006.

4. Flynn, John T. 1932. *God's Gold: The Story of Rockefeller and His Times*. New York: Harcourt Brace.

5. Hays, Samuel P. 1999. *Conservation and the Gospel of Efficiency the Progressive Conservation Movement, 1890–1920*. Pittsburgh: University of Pittsburgh Press.

6. Taylor, Frederick Winslow. 1967. *The Principles of Scientific Management*. New York: Norton.

7. Ibid.

8. Worster, Donald. (N.D.). "John Muir and the Modern Passion for Nature." http://foresthistory.org/Events/Worster%20Lecture.pdf.

9. Henderson, D. 2010. "Paul Samuelson's Prediction for Post World War II." *Library of Economics Liberty*. http://econlog.econlib.org/archives/2010/07/paul_samuelsons.html.

10. Bohanon, C. 2012. "Economic Recovery: Lessons from the Post-World War II Period." Mercatus Center, George Mason University. http://mercatus.org/publication/economic-recovery-lessons-post-world-war-ii-period.

11. The Columbia World of Quotations. 1996. New York: Columbia University Press.

12. *Life.Time.com*. "'Throwaway Living': When Tossing Out Everything Was All the Rage." http://life.time.com/culture/throwaway-living-when-tossing-it-all-was-all-the-rage/#1.

13. Hawkins, Gay. 2006. *The Ethics of Waste: How We Relate to Rubbish*. Lanham, MD: Rowan & Littlefield Publishers.

14. Lindsley, A. 2002. "Profiles in Faith: Dorothy Sayers (1893–1957)." *Knowing and Doing*. http://www.cslewisinstitute.org/webfm_send/445.

15. Sayers, Dorothy L. 1949. *Creed or Chaos?* New York: Harcourt Brace.

16. Häring, Norbert, and Niall Douglas. 2012. *Economists and the Powerful: Convenient Theories, Distorted Facts, Ample Rewards.* London: Anthem Press.

17. Lebow, V. 1955. "Price Competition in 1955." *Journal of Retailing.* http://www.gcafh.org/edlab/Lebow.pdf.

18. Johnson, J. 2013. "Official: Recycling Can Help the Public Feel Good about Plastics." (November 19): http://www.plasticsnews.com/article/20131119/NEWS/131119906/official-recycling-can-help-the-public-feel-good-about-plastics.

19. Packard, Vance. 1960. *The Waste Makers.* New York: D. McKay Co.

20. Ibid.

21. Nordhaus, Ted, and Michael Shellenberger. 2007. *Break Through: From the Death of Environmentalism to the Politics of Possibility.* Boston: Houghton Mifflin.

22. McKenna, J. 2013. "Pope Francis Says Wasting Food Is Like Stealing from the Poor." *The Guardian.* (June): http://www.telegraph.co.uk/news/worldnews/the-pope/10101375/Pope-Francis-says-wasting-food-is-like-stealing-from-the-poor.html.

CHAPTER 4: THE HAPPY CUP FALLACY

1. Drucker, Peter F. 2001. *The Essential Drucker: Selections from the Management Works of Peter F. Drucker.* New York: HarperBusiness.

2. Container Recycling Institute. N.D. "Refillable Bottles: The Decline of Refillable Beverage Bottles in the U.S." http://www.container-recycling.org/index.php/refillable-glass-bottles/53-facts-a-statistics/glass/428-the-decline-of-refillable-beverage-bottles-in-the-us.

3. Google News. N.D. "Litterbug Originated on a Poster in New York in 1947." https://news.google.com/newspapers?nid=1915&dat=19981025&id=Y_dGAAAAIBAJ&sjid=I_gMAAAAIBAJ&pg=5806,5589316&hl=en.

4. Strand, G. N.D. "The Crying Indian." *Orion Magazine.* https://orionmagazine.org/article/the-crying-indian/.

5. Swafford, L. 2015. "Liz Swafford: 'I Want to Be Recycled' Expands." *Dalton Daily Citizen.* (March 11): http://www.daltondailycitizen.com/news/lifestyles/liz-swafford-i-want-to-be-recycled-expands/article_04fd43da-c7a5-11e4-b3b9-97d2679c1e4e.html.

6. Trauht, Erin. 2014. "Take That, GMOs and Pesticides! Organic Industry to Explode to $211 Billion by 2020." http://www.onegreenplanet.org/news/take-that-gmos-and-pesticides-organic-industry-to-explode-to-211-billion-by-2020/.

7. Organization for Economic Cooperation and Development. 2013. "Policies for Bioplastics in the Context of a Bioeconomy." (October 23): http://www.oecd.org/officialdocuments/publicdisplaydocumentpdf/?cote=DSTI/STP/BIO%282013%296/FINAL&docLanguage=En.

8. Mander, J. 1972. "Ecopornography: One Year and Nearly a Billion Dollars Later, Advertising Owns Ecology." *Communication and Arts Magazine*, Vol. 14, No. 2, pp. 45–56.

9. Church, J. 1994. "A Market Solution to Green Marketing: Some Lessons from the Economics of Information." *Louisiana State University Law Center*. http://digitalcommons.law.lsu.edu/cgi/viewcontent.cgi?article=1288&context= faculty_scholarship.

10. Ibid.

11. Bridgeland, J., Putnam, R., and Wofford, H. 2008. "More to Give: Tapping the Talents of the Baby Boomer, Silent, and Greatest Generations." *AARP*. http://assets.aarp.org/rgcenter/general/moretogive.pdf.

12. *Environmental Leader*. 2014. "Consumers Want to Buy Green Brands." (June 6): http://www.environmentalleader.com/2014/06/06/consumers-want-to -buy-green-brands/.

13. TerraChoice. 2011. "Greenwashed: The Truth about 95% of So-Called Green Products." http://terrachoice.com/wp-content/uploads/2011/11 /GreenWashing_Infographic_Nov11.pdf.

14. Verghese, Karli. 2012. *Packaging for Sustainability*. London: Springer.

15. MacKerron, C. 2015. "Waste and Opportunity 2015: Environmental Progress and Challenges in Food, Beverage, and Consumer Goods Packaging." *As You Sow* (January).

16. Elliott, Bobby. 2015. "Waste Management Details Recycling Difficulties." *Resource Recycling*. (February 17): http://resource-recycling.com/node/5692.

17. Ibid.

18. Packaging Strategies. 2014. "Bottled Water Projected to Be the Number One Packaged Drink by 2016." (December 19): http://www.packstrat .com/articles/86173-bottled-water-projected-to-be-the-number-one-packaged -drink-by-2016.

19. Naidenko, O. 2008. "Bottled Water Quality Investigation." *Ewg.com*, October 15. Available at: http://www.ewg.org/research/bottled-water-quality -investigation.

20. Ibid.

21. Environmental Leader. 2010. "Fiji Water Targeted in 'Greenwashing' Class Action Suit." (December 29): http://www.environmentalleader.com/2010/12/29 /fiji-water-targeted-in-greenwashing-class-action-suit/.

22. Ibid.

23. Minter, Adam. 2014. "Why Starbucks Won't Recycle Your Cup." *Bloomberg Review*. (April 7): http://www.bloombergview.com/articles/2014-04-07 /why-starbucks-won-t-recycle-your-cup.

24. Irvine, Martha. 2012. "Young Americans Less Interested in the Environment Than Previous Generations." *Washington Post*. (March 15) http://www.washington post.com/national/health-science/young-americans-less-interested-in-the -environment-than-previous-generations/2012/03/15/gIQAGio1ES_story.html.

25. Kamenetz, Anya. 2011. "The Starbucks Cup Dilemma." *Fast Company*. November. Available at: http://www.fastcompany.com/1693703/starbucks-cup -dilemma.

26. Minter, Adam. 2014. "Why Starbucks Won't Recycle Your Cup." *Bloomberg Review*. (April 7): http://www.bloombergview.com/articles/2014-04-07/why -starbucks-won-t-recycle-your-cup.

27. Geereddy. N. 2011. "Strategic Analysis of Starbucks Corporation." Harvard University. http://scholar.harvard.edu/files/nithingeereddy/files/starbucks _case_analysis.pdf.

28. Kamenetz, Anya. 2011. "The Starbucks Cup Dilemma." *Fast Company*. (November): http://www.fastcompany.com/1693703/starbucks-cup-dilemma.

29. Jewell, J. 2014. "Bottled Water Is the Marketing Trick of the Century." *The Conversation*. http://theconversation.com/bottled-water-is-the-marketing-trick-of -the-century-25842.

CHAPTER 5: BOTTLES GROWN FROM THE SOIL

1. *Institute of Food Technologies*. 2014. "Press Release: U.S. Organic Food Sales Totaled $32.3 B in 2013." (May 15): http://www.ift.org/food-technology /daily-news/2014/may/15/us-organic-food-sales-totaled-$32-b-in-2013.aspx.

2. Hoffman, Beth. 2013. "People Don't Understand What 'Organic' Means, But They Want It Anyway." *Forbes*. (July 17): http://www.forbes.com/sites /bethhoffman/2013/07/17/what-is-organic-anyway/.

3. Dangour AD, SK Dodhia, A Hayter, E Allen, K Lock, and R Uauy. 2009. "Nutritional Quality of Organic Foods: A Systematic Review." *The American Journal of Clinical Nutrition*. 90 (3): 680–5.

4. Lott, Melissa. 2011. "10 Calories in, 1 Calorie Out—The Energy We Spend on Food." *Scientific American*. (August 11): http://blogs.scientificamerican.com /plugged-in/2011/08/11/10-calories-in-1-calorie-out-the-energy-we-spend-on-food/.

5. Gilbert, Natasha. 2012. "One-Third of Our Greenhouse Gas Emissions Come from Agriculture." *Nature: International Weekly Journal of Science*. (October 31): http://www.nature.com/news/one-third-of-our-greenhouse-gas -emissions-come-from-agriculture-1.11708.

6. *International Energy Agency*. 2012. "World Energy Outlook 2012." http:// www.iea.org/publications/freepublications/publication/english.pdf.

7. Stein, M., and Malik, N. 2010. "Just One Word: Bioplastics." *Wall Street Journal*. (October 18): http://www.wsj.com/articles/SB1000142405274870398 9304575504141785646492.

8. Jha, Alok. 2008. "Cheap Way to 'Split Water' Could Lead to Abundant Clean Fuel." *The Guardian*. (July 31): http://www.theguardian.com/environment /2008/jul/31/energyefficiency.energy.

9. Environmental Protection Agency (EPA). 2009. "Draft Regulatory Impact Analysis: Changes to Renewable Fuel Standard Program." http://www.epa.gov /oms/renewablefuels/420d09001.pdf.

10. Mahon, Paul. 2013. *Feeding Frenzy: The New Politics of Food*. London: Profile Books.

11. Conca, J. 2014. "It's Final—Corn Ethanol Is of No Use." *Forbes*. (April 20): http://www.forbes.com/sites/jamesconca/2014/04/20/its-final-corn-ethanol-is-of-no-use/.

12. Miller, H. 2011. "The Great Fuel Fail: Ethanol from Corn." *The Guardian*. (May 12): http://www.theguardian.com/commentisfree/cifamerica/2011/may/12/ethanol-subsidy-usda.

13. Newman, J. 2014. "U.S. Corn Prices Fall to Five-Year Low on Higher-than-Expected Supplies." *The Wall Street Journal*. (September 30): http://www.wsj.com/articles/u-s-corn-stockpiles-rise-after-sharp-increase-in-output-1412093706.

14. Iowa State University. 2015. "Estimated Costs of Crop Production in Iowa—2015." http://www.extension.iastate.edu/agdm/crops/pdf/a1-20.pdf.

15. Gunders, D. 2012. "Wasted: How America Is Losing Up to 40 Percent of Its Food from Farm to Fork to Landfill." National Resource Defense Council. http://www.nrdc.org/food/files/wasted-food-ip.pdf.

16. Laan, Tara, Ronald Steenblik, and Todd Alexander Litman. 2009. *Biofuels—At What Cost? Government Support for Ethanol and Biodiesel in Canada*. Winnipeg: International Institute for Sustainable Development.

17. Miller, H. 2011. "The Great Fuel Fail: Ethanol from Corn." *The Guardian*. (May 12): http://www.theguardian.com/commentisfree/cifamerica/2011/may/12/ethanol-subsidy-usda.

18. *Scientific American*. "The Environmental Impact of Corn-Based Plastics." (July 1): http://www.scientificamerican.com/article/environmental-impact-of-corn-based-plastics/.

19. Pacific Institute. "Bottled Water and Energy Fact Sheet." http://pacinst.org/publication/bottled-water-and-energy-a-fact-sheet/.

20. *Scientific American*. "The Environmental Impact of Corn-Based Plastics." (July 1): http://www.scientificamerican.com/article/environmental-impact-of-corn-based-plastics/.

21. Ibid.

22. Ibid.

23. Pacific Institute. "Bottled Water and Energy Fact Sheet." http://pacinst.org/publication/bottled-water-and-energy-a-fact-sheet/.

24. Atlas, Deborah. 2012. "Plastic Bottles from Plants: Step Forward or Spin Marketing?" *Sierra Club Green Home*. (January 6): http://www.scgh.com/go-green/food/odwalla-plant-bottle/.

25. Ibid.

26. O'Connor, M. C. 2011. "Compostable or Recyclable? Why Bioplastics Are Causing an Environmental Headache." *Earth Island*. (July 6): http://www.alternet.org/story/151543/compostable_or_recyclable_why_bioplastics_are_causing_an_environmental_headache.

27. Bozell, J. 2004. "The Use of Renewable Feedstocks for the Feedstocks for the Production of Chemicals and Production of Chemicals and Material Materials—A Brief Overview of Concepts." *NREL.* http://www.nrel.gov/docs/gen/fy04/36831f.pdf.

28. Mooney, B. 2009. "The Second Green Revolution? Production of Plant-Based Biodegradable Plastics." *Biochem J.* 418, 219–232 (Printed in Great Britain) doi:10.1042/BJ20081769.

29. Kapdan, L. and Kargi, F. 2006. "Biohydrogen Production from Waste Materials." *Enzyme and Microbial Technology,* 38, 569582.

CHAPTER 6: IT'S IN OUR BLOOD

1. Hoferichter, A. 2015. "Remaking the Way We Make Things." *Siemens.com.* http://www.siemens.com/innovation/en/home/pictures-of-the-future/research-and-management/materials-science-and-processing-interview-braungart.html.

2. Freden, J. 2014. "The Swedish Recycling Revolution." *Sweden.se.* (November): https://sweden.se/nature/the-swedish-recycling-revolution/.

3. Nuthall, K. 2015. "EU Commission Revives Circular Economy Plans." (March 9): http://www.europeanplasticsnews.com/subscriber/newsmail.html?id=5684&cat=1.

4. Wachholz, C. 2015. "EU Needs to 'Walk the Talk' on Delivering Circular Economy." *The Parliament.* (March): https://www.theparliamentmagazine.eu/articles/opinion/eu-needs-walk-talk-delivering-circular-economy.

5. *The Economist.* 2007. "The Truth about Recycling." (June 7): http://www.economist.com/node/9249262.

6. *Waste Management World.* 2005. "Recycling Worldwide." http://www.waste-management-world.com/articles/2005/07/recycling-worldwide.html.

7. *The New Age.* 2015. "Germany Threatens Coca Cola with Tax to Deter Disposable Bottle Use." (March 1): http://www.thenewage.co.za/153091-1020-53-Germany_threatens_CocaCola_with_tax_to_deter_disposable_bottle_use.

8. PlastEurope. "U.K. Recycling." http://www.plasteurope.com/news/UK_RECYCLING_t226990.

9. McCurry, J. 2011. "Japan Streets Ahead in Global Plastic Recycling Race." *The Guardian.* (December 29): http://www.theguardian.com/environment/2011/dec/29/japan-leads-field-plastic-recycling.

10. Kanthor, R. 2014. "Global Plastics Recycling Industry Seeing an Impact from the 'Green Fence'." http://www.plasticsnews.com/article/20140421/NEWS/140429996/global-plastics-recycling-industry-seeing-an-impact-from-the-green#.

11. *CCTV America.* 2014. "Challenges Face China's Emerging Waste Recycling Businesses." (December 7): http://www.cctv-america.com/2014/12/07/chinas-emerging-waste-recycling-business-meets-challenges.

12. Tay, H. F. 2015. "Dropping Oil Price Boosts China's Recycling Effort, with Scrap Plastic Industry Feeling the Pinch." *ABC Australia*. (January): http://www.abc.net.au/news/2015-01-21/high-oil-price-gives-chinas-environmentalists-a-boost/6031128.

13. *The Economist*. 2014. "The Eight Year Itch." (September 13): http://www.economist.com/news/europe/21616956-centre-right-government-fredrik-reinfeldt-has-been-great-success-yet-voters-may-well.

14. Karlson, R., and Kuznetsova, A. 2007. "Swedish Environmental Policy." *Baltic Environment and Energy*.

15. Ibid.

16. Avfall Sverige. http://www.avfallsverige.se/in-english/.

17. Avfall Sverige. 2012. "A Year in Brief." http://www.avfallsverige.se/fileadmin/uploads/Arbete/2012_in_brief.pdf.

18. Bottlebill.org. "Sweden." http://www.bottlebill.org/legislation/world/sweden.htm.

19. *Raconteur*. 2014. "Time for Return for Bottle Deposits." (March 5): http://raconteur.net/sustainability/time-for-return-of-bottle-deposits.

20. Karlson, R., and Kuznetsova, A. 2007. "Swedish Environmental Policy." *Baltic Environment and Energy*.

21. Milios, L. 2013. "Municipal Waste Management in Sweden." European Environmental Agency.

22. Seltenrich, N. 2013. "Is Incineration Holding Back Recycling?" *The Guardian*. http://www.theguardian.com/environment/2013/aug/29/incineration-recycling-europe-debate-trash.

23. Polling Report.com. http://www.pollingreport.com/prioriti.htm.

24. Viscusi, W., Huber, J. and Bell, J. 2011. "Promoting Recycling: Private Values, Social Norms, and Economic Incentives." *American Economic Review*. 101 (3): 65–70.

25. Rossignol, K. 2014. "The Ways We Waste: The Long Road to Sustainable Waste Management." *Environmental Protection*. (August 21): http://eponline.com/articles/2014/08/21/the-ways-we-waste.aspx.

26. Bottlebill.org. "California." http://www.bottlebill.org/legislation/usa/california.htm.

27. Ibid.

28. Garrison, J. 2012. "Rampant Recycling Fraud Is Draining California Cash." *Los Angeles Times*. (October 7): http://articles.latimes.com/2012/oct/07/local/la-me-can-fraud-20121007.

29. Elliott, B. 2014. "CalRecycle Retools Under Proposed State Budget." (January 16): http://www.resource-recycling.com/node/4511.

30. Bottle Bill Facts. "80% of Covered Beverages Are Recycled vs 23% Non-Covered Beverages." http://bottlebillfacts.com/wp/?incsub_wiki=80-of-covered-beverages-are-recycled-vs-23-non-covered-beverages.

31. Kadleck, Chrissy. 2015. "Michigan Battles to Improve Its 'Woeful' Recycling Rate." *Waste360*. (February 26): http://waste360.com/business /michigan-battles-improve-its-woeful-recycling-rate.

32. *Maui News*. "Audit: 'Flawed Payment System' in Beverage Container Recycling." (November 19): http://www.mauinews.com/page/content.detail/id /579015/Audit---Flawed-payment-system--in-beverage-container-recycling .html?nav=5031.

CHAPTER 7: SOMETHING NEEDS TO BE DONE!

1. *The Economist*. 2015. "Pocket World Figures."

2. Buffington, Jack, and Ray Peterson. 2013. "Defining a Closed-Loop U.S. Aluminum Can Supply Chain Through Technical Design and Supply Chain Innovation." *The Journal of The Minerals, Metals & Materials Society* (TMS). 65 (8): 941–950.

3. Ibid.

4. Park, A. 2014. "A Frightening Field Guide to Common Plastics." *Mother Jones*. (March 3): http://www.motherjones.com/environment/2014/03/guide -estrogen-common-plastics-bpa.

5. Kavanaugh. C. 2015. "Researchers Focus on Great Lakes Pollution." *Plastics News*. (March 15): http://www.plasticsnews.com/article/20150319 /NEWS/150319902/researchers-focus-on-great-lakes-pollution.

6. United Nations Environment Programme. "Distribution of Marine Litter." http://www.unep.org/regionalseas/marinelitter/about/distribution/.

7. Aldred, J. 2014. "Plastic Waste in the Thames Will Devastate Marine Life, Report Warns." *The Guardian* (January 2): http://www.theguardian.com /environment/2014/jan/02/plastic-waste-thames-marine-life-report.

8. Yang CZ, SI Yaniger, VC Jordan, DJ Klein, and GD Bittner. 2011. "Most Plastic Products Release Estrogenic Chemicals: A Potential Health Problem That Can Be Solved." *Environmental Health Perspectives*. 119 (7): 989–96.

CHAPTER 8: SOLUTIONS—EXTENDING THE SUPPLY CHAIN TO NATURE

1. Bell, Viktor. 2011. "Extended Responsibility." *Recycling Today*. (December 14): http://www.recyclingtoday.com/rt1211-extended-producer-responsiblity .aspx.

2. Cooper, Ben. 2015. "Sustainability in Drinks—Much Work to do on Packaging in the US." *Just Drinks*. (February 26): http://www.just-drinks.com/analysis /sustainability-in-drinks-much-work-to-do-on-packaging-in-the-us_id116266.aspx.

3. Ibid.

CHAPTER 9: 2020: THE HAPPY CUP REALITY

1. *Business Week*. 1998. "Steve Jobs: 'There's Sanity Returning.'" http://www .businessweek.com/1998/21/b3579165.htm.

Index

Aerogel, 120

Aluminum: chemical properties, 29; origins, 28–29; recycling rate, 30; recycling yield, 30, 110–111; secondary use, 30

Asia: growth in beverage use, 36; recycling programs, 88–90

Berner, Robert, 21

Beverages: U.S. consumption, 2; U.S. container use per capita, 36; worldwide growth, 36

BinCam, 83

Biodegradability, 78

Biomass: bioplastic categories, 78–79; replacement for fossil fuels, 71–72; statistics, 71–72

Bottle bill: program, 2, 3, 12; lack of mandated, 27; states with and without, 30

Bottled water: marketing campaigns, 62, 65; municipal water versus, 62; origins, 61

Braungart, Michael, 83–84, 95, 101, 117

California recycling program, 99–100

Carbon abundance, 21

China: Green Fence Program, 88; recycling challenges, 89

Coca Cola: Dasani brand campaign, 63; in Europe, 87; pant bottle 75–76; polyethylene furanoate (PEF), 77; recycling program, 34, 77

Coffee Cups: design, 34–35; lawsuits, 34–35;waste statistics, 64

Composting, 74–75

Container Recycling Institute, 59, 139

Coors Brewing Company, 28–29, 137

Creative convenience, 48

Crowdsourcing, 118–119

Diamandis, Peter, 84

Disruptive innovation, 98, 100–102, 132, 137–139

Downcycling, 85

Eisenhower, Dwight, 54

Environmental Defense Fund, 55

Environmentalism: advocacy groups, 58–59; versus conservation, 43–44; history in U.S., 99; *The Death of Environmentalism,* 50; market research, 56–57

Europe: versus America, 85–86; definition of zero waste, 85; EU Directives, 84; Green dot program, 87, 129; recycling rates, 84–85; waste to energy programs, 84–86, 98

Extended Producer Responsibility Programs (EPR), 87, 129–130

Fiji Water, 62–63
Fossil fuels, 25, 69
Frankenstein, 17
Frankenstein Bottles, 18, 36, 81

Glass: origins in beverage packaging, 26; production process, 26; recycling rate, 27; returnable bottles, 26–27; secondary uses, 27–28
Green chemistry, 116
Green fatigue, 88, 95
Green marketing, 56
Greenpeace, 55
Greenwashing, 57–58, 76

Hagens, Nate, 25
Happy Cup Fallacy, 60–61
Harrelson, Lowell, 9
Hjalmarsson, Pelle, 15

Ickes, John, 25
Industrial Revolution, 24

Jeffery, Kim, 16, 130

Keep America Beautiful, 55
Keurig K-Cups, 34, 35–36

Landes, David, 24
de Lavoisier, Antoine-Laurent, 22

Lebow, Victor, 47
Lignocellulose, 79–81

Malthus, Thomas, 23
Man as Material Scientist 1.0 and 2.0, 18–19, 69, 77, 79, 82, 118
Market-based extended producer responsibility program (M-EPR), 128–133
Material Genome Project, 116–117
Material Scientist 3.0 (supercomputers), 114–116, 119
McDonalds, 48
Michigan recycling program, 100
Minter, Adam, 89
Mobro 4000, 9
Muir, John, 43–44

Napoleon, 25
Nature as Material Scientist 1.0, 18–21, 118
Nature Conservancy, 55
Nature Works, 73
Nestlé Waters, 34, 62, 130
Nocera, Daniel, 71
Nordhaus, Ted, 50
Novelis, 30

Oil: age of, 24; statistics, 69–70
Organics market, 56, 68

Packaging future, 117, 121
Packard, Vance, 39, 46–47,
Perrier, 61
Photosynthesis, 21
Pinchot, Gifford, 43
Plant bottle, 69, 75–76
Plastic: health impact, 111–112; history, 31, 68–69; ocean/wildlife impact, 19, 111–112
Pollan, Michael, 70
Polyethylene furanoate (PEF), 77

Polyethylene terephthalate (PET): design, 32–33, 63; history, 31; recycling rate, 33
Polyhydroxyalkanoate (PHA), 77, 80–81
Polylactic acid (PLA), 73–75
Pope Francis, 51
Prize inducement competition, 118

Recycling: curbside, 3; origins in U.S., 12; market value, 122; versus reuse, 85, 122; single stream, 3; U.S. rates, 1, 123–127
Recycling myth, 2, 123
Recycling reinvented, 129
Republic Services, 11
Rockefeller, John D., 31, 40–42,
Roosevelt, Theodore, 42–43

Sayers, Dorothy, 46
Schumpeter, Joseph, 100
Scientific Management, 42
Senge, Peter, 60
Shellenberger, Michael, 50
Shelley, Mary, 17
Sierra Club, 44, 55, 58
Standard Oil, 42
Standard Packaging, 48
Starbucks: branding challenges, 63–64; origin of cup, 64; recycling program, 35
Sustainability tax, 58
Sweden: versus America, 98: Avfall Sverige, 92; culture, 89–92, 95–96: Förpacknings- och Tidningsinsamlingen (FTI), 93, 129; Lagom, 90: recycling

statistics, 94, 97–98; ReturPak, 94–96; waste to energy program, 98; zero waste definition, 84

Taylor, Frederick, 42
3D Printing, 117–118, 120–121, 140
Throwaway society, 44–45
Tyson, Neil deGrasse, 20

United States: compared to Sweden, 98; environmental history, 99; recycling perceptions, 39; recycling and waste perceptions, 40–41, 99, 103

Voluntary Refund-Deposit System, 122–128

Walmart, 58
Waste: culture of, 44–47; reduction of, 133–135; value in U.S. landfills, 18, 107
Waste Management, 11–12, 60
World War II, 44–45, 68
Wrigley, Tony, 24
Wyeth, Nathaniel, 31

X Prize, 49, 121

Zero waste: Europe definition of, 84; San Francisco's Department of the Environment and, 76; society, 2; and sustainability, 85; and waste to energy programs, 11; and Waste Management, 60

About the Author

JACK BUFFINGTON is a post-doctoral research fellow at the Royal Institute of Technology in Stockholm, Sweden, and is responsible for brewery logistics at MillerCoors, the second largest beer manufacturer in the United States. His research is focused on a cutting edge approach to innovation, science, biotechnology, and the supply chain to improve the environment while growing the economy. Jack obtained a Ph.D. in Marketing and Supply Chain with a focus on innovation from the Lulea University of Technology in Lulea, Sweden. Buffington has also been an adjunct professor at the University of Denver and serves on various professional supply chain boards and associations. Jack also authored the books *An Easy Out: Corporate America's Addiction to Outsourcing* (2007); *The Death of Management: Restoring Value in the US Economy* (2009); *Progress, Technology and Seven Billion People* (2010), and *Frictionless Markets: The 21st-Century Supply Chain* (2015), with distribution markets across the United States, as well as internationally. Jack is intimately involved in the global economy through his writings and key contacts in Asia and Europe, in particular.

.